Living in the Danger Zone

Living in the Danger Zone

Realities about Hurricanes

Bill and Fran Marscher

Writer's Showcase
San Jose New York Lincoln Shanghai

Living in the Danger Zone
Realities about Hurricanes

Writer's Showcase
an imprint of iUniverse.com, Inc.

For information address:
iUniverse.com, Inc.
5220 S 16th, Ste. 200
Lincoln, NE 68512
www.iuniverse.com

ISBN: 0-595-17042-0

Printed in the United States of America

Contents

Preface

In the fall of 1998, we had almost finished drafting *Living in the Danger Zone* when Hurricane Mitch walloped home this significant truth: Every hurricane is a potential killer.

In the late summer of 1999, we were working on revisions when Hurricane Floyd demonstrated another truth: Growth on the coasts from Maine to Texas guarantees massive disruptions in the future as millions of residents evacuate to get out of the way of the hurricanes we know will come.

Although Mitch was not a particularly wretched storm when he struck Central America, he inflicted wretched damage. He had been first a simple tropical wave, then a tropical storm, then a full-blown hurricane, and then a Category 5 hurricane with sustained winds of 180 mph. After ravaging islands in the Caribbean, Mitch dropped his wind speeds—but hung onto the ocean of rain in his clouds. Then he meandered slowly into the rugged, rural mountains of Honduras, Nicaragua and Guatemala, dumping 35 inches of water in three days.

Mitch swept away entire villages and a vast region's network of roads and bridges. In his wake, corpses floated in the rivers and lay beneath the debris, including the slopes of hills that had slumped into the valleys. Although no one knows for sure, officials believe Mitch's casualties included 9,000 deaths and 9,000 missing.

"Everything we own has been obliterated," said Sergio Santeltz, who lost 21 relatives to the terrifying avalanches of mud and water. "My family has been lost to the earth."[1]

The International Red Cross hurried to Central America before Mitch disappeared off the map. They brought compassion, creativity and know-how. Still, for weeks afterward, Mitch's survivors faced starvation and deadly disease, so cut off were they from the resources that poured by airlift into their countries.

Eleven months after Mitch, the threat of Category 4 Hurricane Floyd as he traveled parallel the Florida, Georgia and South Carolina coastlines dictated the largest evacuation ever in the United States. An estimated 2.6 million, more or less, packed their insurance papers, their family photo albums and their pets and fled from their homes, creating a three-state traffic jam within a 24-hour period. The critiques of the evacuation began immediately.

In North Carolina, another half a million were ordered to evacuate before Floyd slammed ashore at Cape Fear, creating what officials called the worst environmental, agricultural and human disaster in the history of the state.

So, while Floridians, Georgians and South Carolinians wrestled with the question of what to do next time to ensure a less exasperating evacuation, North Carolinians struggled under hazardous, destructive flooding deeper and more widespread than anyone alive could remember.

No ready answers

Coincidentally, Hurricane Floyd uncannily resembled the Great Sea Island Storm of 1893, which tracked the Florida and Georgia coast before going ashore at Savannah. We two coastal South Carolinians watched Floyd with great respect. When the evacuation order came for our community, we dutifully boarded up the house, packed the computer disks and the family photos and headed inland. And we experienced quite personally a piece of what we had only read about and interviewed about and written about before—the tension of sitting for hours in a line of traffic barely crawling toward high ground, with a then Category 4 hurricane aiming in our direction.

The Hurricane Floyd trauma that hit the whole Southeast region in various ways moved The Weather Channel and the American Meteorological Society to pull hurricane experts into a workshop for brainstorming in June 2000. They asked: "What is wrong with this picture?"

In a report published three months later, they concluded: "Today's hurricane warning and response system is more capable than ever before, but it is inadequate to avert a major catastrophe with its attendant large loss of life, enormous property destruction and business disruption, and long recovery times. One of every five people in the United is at direct risk of hurricane impacts, and the number is growing daily. During the last decade or two we have been very fortunate. Considering the large increases in population and development in hurricane prone areas, storms of great magnitude similar to those experienced in the past, striking the same region now, would cause much greater losses in life and property.

"Business-as-usual approaches such as incremental improvements in prediction skill and modification in present evacuation procedures must be augmented by substantial advances in hurricane preparedness, forecasting and response strategies and by more effective coordination. To substantially reduce future hurricane losses will require concerted action on the part of all levels of government, as well as private enterprise and the general public."[2]

That high numbers of Americans might drown in a hurricane in this century seems unthinkable, but the unthinkable, according to the best information available, could be on the verge of reality. In addition, mammoth property losses loom on coastal horizons.

It's not as if no one is working on the problems described in the report mentioned above, which are the problems we describe in this book. Those who understand best the high-stakes gamble in the hurricane-vulnerable regions inform and warn the public as best they can. But the professionals have a limited ability to make a difference. The search for solutions leads straight to those living in the danger zone.

Together, the many sources of information we studied about hurricane risks on the Atlantic and Gulf coasts point toward a future filled with more hurricane drama, and tragedy.

It has taken a community of experts to raise this book. Our gratitude goes to dozens of sources who became friends, plus dozens of people we never met, including those who recorded their personal impressions of hurricanes in all their harrowing detail. Our deep appreciation goes in large measure to scholars and practical planners who studied and built a body of facts and conclusions about what hurricanes do to the atmosphere and to human populations. William Winn, director of emergency services in Beaufort County, South Carolina and a member of the Federal Emergency Management Agency's Liaison Team, encouraged us and taught us patiently. Robert Collins, the state of Florida's hurricane planner, shared with us much of his valuable insight. Jerry Jarrell, former director of the National Hurricane Center, worked to help us understand the intricacies and gutsiness of forecasting. Disaster-relief experts Pat Goodale and Michael Logan of the American Red Cross gave us a keen appreciation for the importance of trained volunteers who hold hurricane victims' hands and give them a hand-up. Dr. Peter Sparks and his team of wind engineers at Clemson University showed us what it takes for a family's home to hang tight when one of these dangerous, powerful monsters is raging all around. Charles Watson Jr. of Watson Technical Consulting walked us through the process of applying state-of-the-art-computing technology to the science of hurricane damage prediction. With the counsel of Charles "Chuck" Gregg of the Georgia Emergency Management Agency and Stan McKinney of the South Carolina Emergency Planning Division, we were provided access to many other experts.

Finally, retired magazine editor Don McKinney encouraged us to continue working beyond the first draft to make the subject more accessible for readers.

Bill and Fran Marscher

Chapter 1

Nature on the rampage

Every hurricane a potential killer

"God moves in mysterious ways
His wonders to perform;
He plants his footsteps in the sea,
And rides upon the storm.

William Cowper

Blame the deadly Atlantic hurricanes on high pressures over Germany. Blame them on the high plateaus of the Sahara, which further heat the hot dry air flowing out of Europe. Blame these monsters from the sea on the easterly trade winds, which help move them across the ocean. Blame the ferocious Gulf hurricanes on the Caribbean's torrid waters.

Better still, blame these killer storms themselves for their power and their paths. Once hatched, these things take on lives of their own.

Along with the typhoons of the northwest Pacific Ocean and the cyclones of the south Pacific and Indian Ocean, hurricanes are the world's climatic monsters. Harness one day's release of energy from only one average hurricane, and you could supply electricity to the entire country for six months.[3] A spinning tornado may have a higher wind speed, but it is short-lived and narrowly confined. Full-grown hurricanes churn air and water into a turbulent mass that reaches from the surface of the ocean to the stratosphere 50,000 feet up. They dominate weather and waves over thousands of square miles.

These gigantic weather phenomena are born over sun-warmed ocean water in the tropics. Heated air rises, cools, spreads out in the upper atmosphere and begins to sink in a roughly circular pattern. The falling cooled air is then sucked back up into the low-pressure core, where it heats and spirals upward. The spin of the earth twirls the convection system clockwise south of the equator and counterclockwise north of it.

So the mass of air is swirling around and also is being pumped by the internal, vertical exchange of hot and cool air. Meantime, various air currents, themselves influenced by differences in pressure and temperatures over a deep and wide portion of the atmosphere, determine the storm's shape, its intensity, its size, its forward speed and its track.

The storm surge

From the standpoint of loss of lives, the greatest hurricane catastrophes in history have occurred in connection with the storm surge, called in 1893 the "storm wave," also called a "tidal wave" in various accounts of hurricanes past. In general, the lower the air pressure in the center core, called now the "eye," the higher the wind speed in the "eye wall" surrounding the eye, and the higher the dome of water that races onto the open coast.

The myth that the hurricane's low-pressure core pulls a churning dome of water up out of the ocean and drags it along is not exactly right. That accounts for only one to two feet of the storm surge. Most of it builds up because of high winds whipping the seas of the vast ocean, pushing and piling them into a moving mound of water.[4]

When the hurricane's counterclockwise winds hurl that mass of water onto the shore, it is a killer force, most dangerous to whoever or whatever is caught in its right forward quadrant. It mounds highest where the continental shelf is shallowest. Storm surges of more than 20 feet above mean sea level, topped by hurricane-powered waves of 20 to 30 feet, have barreled ashore. The worst of all possible times for such a strike is on a high tide, especially in regions of great tidal range.

Although every hurricane has the power to kill and destroy, some carry more danger and produce more damage than others. On the basis of wind velocity, barometric pressure and height of the storm surge, the Saffir-Simpson Hurricane Intensity Scale classifies hurricanes and assigns them one of five categories. The National Hurricane Center provides descriptions of the potential damage for each category. For example, a Category 1 will damage unanchored mobile homes, trees and signs; cause minor flooding in low-lying lands and rip small boats from their moorings. A Category 2 will rip into roofs, doors and windows; crack limbs and knock down small trees; shake up docks and boats. A Category 3 will destroy mobile homes, produce significant structural damage to many buildings, residential and commercial, and cause substantial flooding. A Category 4 will demolish many structures, rip trees out of the ground and devastate areas near waterways. A Category 5 will make debris out of the development in its path—and is capable of dramatically altering the landscape. (For more on the Saffir-Simpson Scale, please see the appendix.)

In addition to putting hurricanes into categories, scientists name them. In some communities just uttering a single name conjures up indelible recognition of devastation and misery—Camille, Agnes, Hugo, Andrew. (For more on hurricanes' names, please see the appendix.)

That almost 20 percent of Americans would voluntarily live, work and play where these destructive phenomena barrel ashore every year seems peculiar. That the pace at which Americans move to the vulnerable shores would exceed by 50 percent the pace at which they move elsewhere in the 1990s seems puzzling. That Americans would invest in their collective gross domestic product (the total annual value of goods and services produced) at a rate of 20 percent faster per year in the hurricane-vulnerable regions than in the rest of the country is almost bizarre. [5]

Nevertheless, such peculiar, puzzling and bizarre behavior occurs pell-mell at the beginning of the 21st century. It's as if there is no history. And it's as if there's no tomorrow.

Ships' reports

Long ago, the chief sources of pre-landfall hurricane news came from ships' captains whose sailing vessels or steamers unsuspectingly were caught in them. By the 19th century, the seamen could collect data on temperatures, air pressure, wind direction and wind speed wherever they happened to be at the time. Those who lived through storms at sea to tell their stories could report how rough their voyages became, and how deluges of seawater washed their colleagues overboard. They could tell very little about the size, the intensity, the pace or the movement of these ocean-born storms, of course, and they could not deliver their reports until they came into port. They furnished useful information but not in a timely enough fashion for coastal dwellers to be able to flee from the dangerous places.

Historians have pieced together tales of long-ago storms in the Atlantic, starting with a "September gale" that Christopher Columbus ran into on his second voyage to the New World in 1494. Columbus had been forewarned, he said afterward, of a cyclone by the appearance of a "sea monster" on the surface of the ocean. He and his small fleet rode out the hurricane near Hispaniola (now San Salvador in the West Indies).

In the 115 years between Columbus' discovery of land in the Western Hemisphere and the permanent English settlement of what is now the

United States, a lot of hurricanes bounced and battered the Spaniards in their sailing ships. In 1565 the Spanish and French fought bitterly over territory near St. Augustine, Florida until what is now assumed to have been a hurricane sent the French fleet to the bottom of the ocean.[6]

Hurricanes continued to hamper exploration, treasure seeking and settlement. In the more than 200 years before systematic hurricane study began, as New World adventurers made their tentative way in the hurricane belt, they got no timely notice of the violent attacks that came upon them. How many ships were beaten to pieces and how many sailors and passengers drowned as the Europeans tried to sail to and from the Americas in those years remains one of the ocean's great secrets.

'Marks of it will remain this hundred years'

Although hurricanes strike most often along the shores of the Southeast Atlantic or the Gulf of Mexico, they have wrecked New England in the past and will likely do it again. One Yankee wrote this about one of them in 1635: "This year, the 14th or 15th of August (being Saturday) was such a mighty storm of wind and rain as none living in these parts (New England), either English or Indians, ever saw. Being like, for the time it continued, to those hurricanes and typhoons that writers make mention of in the Indies. It began in the morning a little before day, and grew not by degrees but came with violence in the beginning, to the great amazement of many.

"It blew down sundry houses and uncovered others. Divers vessels were lost at sea and many more in extreme danger. It caused the sea to swell to the south wind of this place above 20 foot right up and down, and made many of the Indians to climb into trees for their safety....The signs and marks of it will remain this hundred years in these parts where it was sorest."[7]

A little more than a century later, Benjamin Franklin figured out that these hurricanes do not just spring up on the spot but travel. He had planned to watch the scheduled eclipse of the moon from Philadelphia one evening in 1743 but was disappointed because storm clouds blocked his view. A Philadelphia paper later made reference to "a violent Gust of Wind and Rain attended with Thunder and Lightning."

Ben Franklin learned later that Bostonians got a good look at the eclipse in a cloud-free sky—before stormy weather arrived there later that night. Noting the difference between the times of the hurricane conditions in the two cities, he concluded that they experienced the same storm, which must have traveled from southwest to northeast against the current of the surface wind.[8]

Another century passed between the time of Franklin's discovery and the beginning of the U.S. government's involvement in weather watching, reporting and warning. Meteorology as a science was in its infancy. The Constitution had dealt with Congress and religion and search and seizure but had not addressed weather. Politicians did not link "public welfare" with study of the atmosphere to predict whether it would rain or snow or blow like a hurricane.

Nation at risk

In the 18th and 19th centuries, it became increasingly obvious that this young country had better pay attention to these destructive storms. More and more lives and property were at risk along the hurricane-vulnerable coasts. And yet, even if the government had made a concerted commitment to meteorology, without the ability to communicate quickly, or to communicate from ship to shore, weather observers lacked basic tools of weather forecasting in the public interest.

In the three decades after the telegraph was patented in 1837, lines expanded rapidly. They crossed the Atlantic Ocean in 1866. In 1895, an Italian produced the first wireless telegraph, and in 1900 radio-voice transmission began.

By 1850, the U.S. Army Signal Service had a network of 150 weather observers scattered across the country. By telegraph, data from these observers clicked its way to Washington, DC, so scientists could construct a daily weather map. The Washington weathermen then telegraphed their conclusions and predictions back out across the country, and regional observers "posted" advisory weather signals as needed. "Posting" meant writing notes on bulletin boards around town, mailing post cards and raising flags on the tops of buildings.

Twenty years later, in the hopes of protecting shipping interests, the U.S. Congress wrote a law directing the Army to learn more about storms affecting the Great Lakes and the Atlantic coastlines. Still, weather reporting at the time was not considered a service necessary for *human* safety. These Signal Service officials were expected to prove their worth—and thus keep their jobs—by listing and describing ships' *cargoes* that, in the absence of the forecasts, might have been destroyed by rough weather.

Weather service in the public interest evolved. The Signal Service could not be right every time with its hurricane forecasts. In those early years it took some grief as well as credit. The Army's military brass wanted to get the weather business out of their jurisdiction. In 1880, the function was turned over to U.S. Department of Agriculture and renamed the Weather Bureau, and the Signal Service workers became civilians. A fuss about whether or not federal funds should be spent on weather stations in the West Indies erupted briefly, but by the end of 1881 the legality of such expenses was established and six stations in the Antilles began reporting.

400 years

Still, in the four centuries that elapsed between the storm that unexpectedly caught Columbus' fleet off Hispaniola in 1494 and the Great Sea Island Storm of August 1893, little had been learned and little communicated to the public about hurricanes. It is doubtful that anybody lobbied at all during those years for more research and more education about what these "cyclones," as they were called, could do to human populations.

For one thing, the numbers of people living in the various danger zones on the Atlantic and Gulf coasts were small. In the eighteen coastal states and the District of Columbia in 1900, the total population was only 32 million. In contrast to Florida's population of nearly 15 million today, only about one half million people lived in all of Florida in 1900.[9]

In addition, those who settled in the coastal states in the nation's first three hundred years had not situated themselves in the most vulnerable places—the oceanfronts, the tidal floodplains, the freshwater swamps. They had no hurricane-risk statistics at their fingertips, as we do today, but they used common sense.

The coastal properties of lowest elevations were teeming with wildlife, including elusive minks and ubiquitous alligators and mosquitoes. Trappers and a few fishermen and farmers used them, but the most hurricane-vulnerable lands were only sparsely settled. Along the shores of the Atlantic and Gulf, especially, the islands were considered undesirable sites for people seeking the good life.

So hurricanes swept across the low lands for many a century, reshaping the inlets, the sand shoals and even the vegetated islands, only rarely harming humans. Few Americans realized that these hurricanes could take lives on land as well as on ships at sea.

Catastrophe strikes

Then, in August 1893, the Great Sea Island Storm, that we now know was a Category 3 hurricane, drowned 2,000 or more. The hardest hit population, the residents of South Carolina's Sea Islands, could not have been more vulnerable. Former slaves and descendants of slaves lived in fragile cabins on flat, swampy lands. The ocean's surf pounded the beaches a day or two ahead of the onslaught, and the alligators probably bellowed, but these coastal dwellers had no other warnings in advance.

They tried to go about their normal business Sunday morning, August 27, 1893, despite the beginnings of what they called "the weather"—scudding cirrus clouds and gusting breezes. By mid-afternoon, limbs were crashing, the water in the creeks and sounds was getting rougher, and it was clear that "the weather" was building up instead of abating. By dusk, fear and dread filled the hearts of the 200,000 coastal folk from Savannah, Georgia to Charleston, South Carolina, and shortly before midnight, the eye of the hurricane came ashore just south of Savannah. The counterclockwise swirl of its high winds shoved the deadly 10 to 13 foot storm surge over the Sea Islands.

In the drenching rain, Savannah's streets became streams. Into the thrashing Okatie River of Southern Beaufort County, the bluffs near the present Sun City Hilton Head crumbled. Beaufort's senator described a tide that should have been high at 10:30 p.m., that was "stationary" at midnight and that then rose two and one half feet and "came in upon us with a sudden rush, eight feet above ordinary spring tide." In Charleston, the winds gusted to 120 mph shortly after midnight.

But it was on the islands—Daufuskie, Hilton Head, Pinckney, St. Helena and Edisto—that the destruction was nearly complete. By daylight Monday, rubble covered what had been modest neighborhoods. Corpses lay in surprising places. The fields of corn were flooded, and the food supplies, including the livestock, were gone.

So isolated were these communities that almost a week passed before help came to them. By Friday, after the Sunday night devastation, inquiries began to pluck the night-of-the-storm stories from the survivors. At first, the newspapers estimated the number of fatalities at 600, then 1,000 and then 1,500. Eventually, estimates of deaths rose to 2,500 or 3,000, counting those who died afterward from injuries, malaria, dehydration and other hurricane-related illnesses.

As the nation learned about the disaster, donations of cash and clothing began to arrive by train. Everything needed had to be produced out of the generosity of outsiders—tools, lumber, nails, pork, disinfectants, everything. Nothing in Americans' experience had prepared them to assist thousands of hurricane victims suddenly left without life's barest necessities. Neither the U.S. Congress nor the S.C. General Assembly appropriated a dime to help.

Fortunately, Clara Barton, president of the American Red Cross, assessed the situation in mid-September and agreed to take on the relief and recovery effort, knowing that she would have to keep a steady appeal going through Northern newspapers to produce the necessary cash and goods. Even she did not estimate the magnitude of the problem, though. She did not realize that she would be trying to stave off the near-starvation of 70,000 people. She did not recognize how much ditch digging and rebuilding would be required. She could not have known how seriously malaria, poor nutrition and dehydration would affect the storm's victims. She almost certainly would not have anticipated that it would take the Red Cross almost a year to put the region on a safe path to recovery.

The Great Island Storm's survivors felt its physical and psychological blows for the rest of their lives—and passed their family's hurricane stories on to their children, who also passed them to their children. Even today, when a hurricane appears on the weather map, the legend of that hurricane rises like an unfriendly ghost among the gnarled limbs of the live oaks in that region.

Chapter 2

---

959 Terrors from the Sea in 111 Years

Then Galveston, and 1935 in the Keys and...

His chariots of wrath the deep thunder clouds form,
And dark is his path on the wings of the storm.

Robert Grant

South Carolina Lowcountry folk took a beating one night in August 1893 like nothing that had ever happened to so many Americans. Massive hurricane fatalities were a new phenomenon in this country.

To this day, never have so many Americans faced such utter famine for so long.

Later, weather-watchers would call the Great Sea Island Storm the "forgotten hurricane"—"forgotten" perhaps because it wreaked its destruction on remote islands and swamps and because most of its victims were black, former slaves and descendants of slaves. But in October of that same year, another little remembered hurricane drowned 1,500 in Louisiana.

Galveston, deadliest ever

And in 1900, only seven years after the two deadly blows to South Carolina and Louisiana a powerful hurricane roared out of the Gulf of Mexico onto Galveston Island, Texas. To this day, no other storm and no earthquake, no flood, no fire, no tornado, no epidemic of disease and no volcano has killed so many Americans so fast as the great Galveston Hurricane. Almost certainly as many as 6,000, and possibly as many as 8,000, drowned in the turbulent flood waters, were crushed by collapsing buildings and chimneys or were battered by the flying roof tiles and the charging wharf pilings and crashing trees ripped from the places they had stood for years. As every storm-struck community does, Galveston suffered during and after the hurricane in its own way.

In contrast to the way former slaves and their descendants lived on the low-lying islands of South Carolina in 1893, Galveston Islanders lived in "high cotton" at the turn of the century. A strip of sand twenty-eight miles long and one to three miles wide, Galveston Island had attracted gentlefolk from upcountry agricultural towns to summer there in comfort and luxury. Soft breezes from the Gulf and crisp, white sands along the shore had inspired growth of a lively resort colony of splendid homes ornamented by palms and oaks and oleanders that bloomed profusely in the heat of July and August.

Granite jetties installed to force scouring of the depths of the harbor had helped Galveston become the main port of entry for vast Southern and Western farmlands. *Death from the Sea* by Herbert Molloy Mason Jr. tells the Galveston drama effectively.

"As the summer of 1900 drew to a close, Galveston was never more beautiful, never more prosperous....

"Great coal-burning cargo ships, proud square-rigged barkentines with billowing canvas, and low-lying coastal schooners paraded constantly along the ocean roads leading to Galveston, keeping the miles of wharves filled to capacity. Twelve hundred ships a year on the average entered and cleared with more than $300 million in cargo. Out flowed nearly 70 percent of the nation's cotton export, and grain, flour, breadstuffs, zinc, ore, sawn lumber, cottonseed oil, beef, hogs and dairy products. In came beet sugar, jute butts, sisal, cement, coffee and chemicals."

…"Feeding and fed by these (many, many industries) were 37,789 people, most of whom claimed that they lived in the finest city in the state, if not along the entire Gulf Coast. Galveston was certainly the most beautiful."[10]

In late August 1900, over the ocean's summer-sun warmed waters, about 1,000 miles west of Senegal, an "easterly wave" began shaping up into what would become Galveston's nemesis. On August 30 ships' crews trying to transport cargo 600 miles east of Puerto Rico were the first men to feel the fury of the whirling mass of wind and water. By September 5 it was swamping Havana, Cuba, and aiming toward Key West, Florida.

"This hurricane, nameless and newly fueled by warm offshore waters, swelled until it measured four hundred miles from one ragged edge to the other, the gyrating arms encompassing more than 150,000 square miles of heavy rains and tearing winds."

It had slapped at Trinidad, and it now slapped at ships in the Atlantic, the Florida Straits and the Gulf of Mexico, and teased Tampa, Florida; New Orleans, Louisiana and even Atlanta, Georgia, with its widespread power. Coastal people of Louisiana, Alabama and Mississippi, remembering a previous, little known hurricane that had drowned hundreds in their flat, low-lying communities in October 1893, quaked at the thought of landfall near them.

Abruptly, the hurricane quit its northward path, picked up speed and headed straight toward Galveston Island.

Molloy gives the picture of the city: "On Friday morning, September 7, 1900, the storm-warning flags…run up over Galveston…snapped smartly in the breeze."[11]

To evacuate or not

Galvestonians who telephoned the Weather Bureau asking what was coming and what they should do about it were told accurately that no one knew how bad the hurricane would be—but that it likely would be troublesome and that they should head for high ground if possible.

But should they try to catch the train to the mainland? Individual households made their decisions on their own. Only a handful fled.

By 10 a.m. Saturday, rain was falling, wind was blowing, the surf was beginning to roll into the streets, and salt spray flew not only over the wharves but over the picket fences and then onto the porches of the Galveston houses. Sightseers rushed to the beachfront.

"The holiday atmosphere among the curiosity-seekers was not apparently dulled by the sight of the waves wrenching loose the stairways and platforms leading from the beach to the bathhouses....None seemed to realize that the storm was not an entertainment, but an executioner."[12]

By 3 p.m. water covered the city. By 5 p.m. the wind reached 74 mph in the city that was the pride of the Gulf coast. Before 5:30 p.m., the wind gauge hit 84 mph, then blew loose.

A sturdier porch floated by

The water continued to rise. Anthony Credo chopped holes in the floor of his parlor, hoping that by letting the water in, he would prevent it from floating the house away. His strategy failed. After darkness fell, the storm surge tore the house from its foundation, and the family crawled out of the attic windows into the deep, rough water. The man, woman and children clung to the roof of the house until they realized the waves were breaking it up. Fortunately, a sturdier porch floated by. It too was bobbing, creaking and cracking, but they climbed onto it and so kept all but one family member alive that night.[13]

The Credo family's experience mirrored that of tens of thousands of others. Altogether, almost 40,000 people spent that harrowing Saturday night on the Texas coast fighting for their lives and watching in horror as their neighbors and children and parents lost theirs. On Sunday morning the corpses lay on the beach, in the streets, under debris. All communication and transportation to the rest of the world was cut off.

By Monday, it was obvious that normal burials would be impossible. The stench of the dead and the dread of disease from the rot of flesh forced Galvestonians to load bodies of their friends and families onto barges, tie ballast to them or tie them together and dump them into the Gulf of Mexico. By Tuesday, the tide and the surf had brought some of the bloated corpses back to the beach.

Not because of its physical size or its power but because of where it hit—and the number of people in its way—the data about this storm still skew the statistical calculations hurricane historians use to compare these complex tropical weather systems.

Similar hurricane terrors, however, have ravaged islands and coastal communities and even towns as far inland as 200 miles from the ocean. Despite the known science of these systems, they behave in quite individual, random, unpredictable fashion. When one considers dangers and destruction, "category" is only part of every picture. What happens to property and populations is influenced by dozens of factors, some of them dependent on the storm itself, others dependent on the lay of the land— and the critical timing and the whimsical angle of landfall.

Texas lore includes many other hurricane tales, including those of the Galveston hurricane of 1915, also a Category 4. By 1915, the city had built a seawall, in places as high at 17 feet, and painstakingly and at great expense had physically raised all of its infrastructure, some of it as much as 13 feet, in an area forty blocks long and from two to twenty blocks wide. This time the loss of life and the destruction were not nearly so severe.

Florida's hurricane history

The state with many more hurricane tales in its history than Texas—more than any other state—is Florida.

In 1928, only six tropical storms formed in the Atlantic, but one of them, a Category 4 hurricane, so filled and whipped Lake Okeechobee that its waters almost met the ocean itself. Close to 2,000 died in that one.

On Labor Day Weekend in 1935, the Florida Keys, on which there is nowhere to hide from wind and water, took one of only two Category 5 hurricanes to strike the continental United States in this century. Imagine sustained winds of more than 155 mph on the slender coral strip of islands south of Miami. One old-timer recalled a 20-foot wall of water that toppled locomotives and passenger cars "like they were nothing." The hurricane crushed his mother under an icebox and washed his father into the mangroves. He became an orphan that day. Four hundred and eight died in that storm.[14]

Long-time Floridians can talk of Hurricane Donna in 1960 and of Hurricane Betsy in 1965 and of other times that the soft Sunshine State, bright with blooming hibiscus, became hard and frightening, dark and deadly.

Destruction from Andrew, which slammed into Homestead south of Miami in 1992, eclipsed everything before it. He was a fierce wasp, an angry bantam rooster, a physically small but nasty-tempered storm.

Floridians knew Andrew was coming. Officials activated their emergency plans. Residents stocked up on canned food and batteries, stowed their bicycles and deck chairs. Some of them boarded up the windows. As many as 750,000 evacuated from their waterfront properties and mobile homes. But the community was hardly braced for what it got.

'Andrew chased Joe'

On Monday morning, August 24, 1992, according to *Andrew! Savagery from the Sea* by the Fort Lauderdale Sun Sentinel, "Joe McCarthy tried to shut Andrew behind the doors of his home.

"From room to room to room to room—through five bedrooms and nearly two hours of terror—Andrew chased Joe and his mother and sister through the home where Joe grew up.

"'The storm would come into another room and we would get that door closed, and then we'd hear windows breaking,' said McCarthy, 29, 'It was sort of like somebody banging on the walls trying to get in.

"'It was a personal kind of thing. Like, 'That storm is not going to come and damage this place. But now you realize there's nothing you can do anyway.'"[15]

By the time Andrew had finished its battering in Florida and taken off toward Louisiana even those who knew Dade County best had a hard time finding their way around the rubble. Houses were in splinters, trees prostrate, streets strewn with soggy sofas and broken toilets. Automobile showrooms, large and small businesses, the Metrozoo's exhibit of 350 exotic birds and much of Homestead Air Force had been wiped out. The 140-mph winds, gusting to 169 mph, and the storm surge reaching 16.9 feet, had done their deeds.

Andrew killed only 39 people that Monday in Florida, but he killed electrical power for 1.4 million and smashed the homes of 250,000 in varying degrees. "Dazed and homeless storm victims stumbled about in search of food, water and shelter," wrote a Fort Lauderdale Sun Sentinel reporter.[16]

Metro-Dade emergency operations manager Kate Hale said grimly: "We knew within hours that south Dade would literally have to be rebuilt from the ground up. It was just a staggering prospect."[17]

Actually, Andrew would have caused more costly anarchy among three times as many victims if he had come ashore fifty miles north of where he did. Instead of attacking the Homestead area of 71,436 homes with a population of 202,036, he could have attacked Fort Lauderdale, a community of 179,304 homes and a population of 614,558.

What dictated Andrew's precise path? Which elements in the atmosphere, or, above it, in the troposphere several miles above the surface of the earth, pushed or pulled Andrew southward?

'Hurricane Alley'

How, North Carolinians might ask, can anyone explain the extraordinary bad luck of experiencing four hurricanes in three years near the end of the 20th century?

Along the state's Outer Banks and other coastal communities, the Great Hurricane of 1899 stands out for its furor against ships and its massive storm surge on Cape Hatteras. In the 1950s, the North Carolina coast became known as 'Hurricane Alley'—a path hurricanes took doggedly for a few years. Hurricane Hazel in 1954 rolled an 18-foot storm surge into Calabash on the highest tide of the year before ramming its torrential rains and 100-mph winds all the way through Virginia and Pennsylvania.[18]

In 1984 Hurricane Diana made national history for wrecking not only both coasts of Florida but also the Carolinas and New England.

However, in recent memory it is the incredible 1990s that showed North Carolinians how vulnerable they are. In July 1996, Hurricane Bertha blasted the state. Clean-up was still under way when Hurricane Fran stuck hard with wind and storm surge, leading to a disaster declaration in 51 counties.

Then, in August 1999, Hurricane Dennis lashed the coast briefly before meandering offshore as a tropical storm and intensifying, then coming back into the state again, this time with sustained 70 mph winds and rain that would not stop. September 1999 brought even worse trouble in the form of Hurricane Floyd, which, in addition to killing 69 on the tropical offshore islands and on the continent, wreaked unspeakable horror and destruction in North Carolina, mostly from rain-driven flooding.

Arbitrary and capricious

On maps, the tracks of hurricanes look like the cat's knitting yarn. (Please see the illustrations for a map showing hurricane tracks.) They form near Africa or in the Caribbean or the Gulf of Mexico and sweep and wobble and curve and curve back and generally move westward and northward. They appear to have minds of their own. They might take this track and might take that one. The jet stream could push the hurricane this way or it might not. The computer models for predicting path often produce different predictions. The computer models that predict changes in power do not always agree with one another.

When a hurricane is traveling near the coast or toward the coast, the unknowns are maddening—to everybody involved. That includes not only the public that could be affected but also officials, businesses, tourists, residents, all of those who go into a "stand-by" mode in potential emergencies. The uncertainties are not going away, however, so we ought to get used to them.

Between 1886 and 1997, 959 tropical cyclones formed in the Atlantic Basin. Some never became hurricanes. Most did not strike the mainland of the United States.

Since 1900, from four to ten "major" hurricanes have struck the continental United States per decade. "Major" means Category 3, 4 or 5. Three "major" hurricanes hit in 1909, again in 1933, again in 1954. Six hurricanes, not all "major" but all with hurricane-strength winds of 74 mph or greater, hit the country in 1916—and again in 1985.

The statisticians and meteorologists can report the numbers. The records are fairly thorough. In retrospect they can say what happened.

They have a harder time explaining the "whys."

Great New England hurricane of 1938

Few New Englanders expect hurricanes to reach *their* shores, especially not at full hurricane force, straight in from the ocean. Hurricanes often either stay at sea their whole lives or wear themselves out along the Southeast or the Gulf Coast before trailing wind and rain to the Mid-Atlantic and Northeast on their way to petering out.

With the exception of the area's disaster preparedness officials, New Yorkers and Bostonians spend little time worrying about hurricane damage or hurricane evacuation or hurricane mitigation or post-hurricane plans. To most, snow is a bigger problem. Most of the population might be shocked speechless to see the heartbreaking photographs taken after the New England hurricane of 1938.

Several days of heavy rain had swollen New England's rivers, washed out sections of the New Haven Railroad and flooded highways before that storm hit. "The hurricane, after skirting the New Jersey shore, struck Long Island and the New England coast with terrifying force about mid-after-noon (September 21), tearing a wide swath from ...New York City on the west to Boston on the east, ripping its way northwards, tearing down trees, buildings, telephone and telegraph poles and wires, and leaving behind a toll of death and destruction seldom, if ever before, equaled."[19]

The storm surge, called at the time a "huge tidal wave," engulfed every-thing in its path, as much as a mile inland, and the Harvard Meteorological Observatory officially recorded the wind at 186 mph.[20]

The New England hurricane not only wrecked houses and downed trees, as hurricanes were known to do in the rural South, it also demolished urban industrial complexes, mazes of trolley lines, cargo ships, tankers, whole factories and train depots. It wrenched an 800,000-gallon oil tank loose in the New Haven Harbor and blew it into a bridge across the Quinnipiac River. Plucky Yankees lucky enough to get some sleep woke the next morning to find everything in disarray, hardly anything in the landscape where it had been the day before. *The New York Times* made an early estimate of $500 million (1938 dollars, equal to about $3.5 billion in 1990 dollars) in damages. The American Red Cross estimated fatalities at 682.[21]

As tough as each bit of destruction was to each individual or business affected, nothing was more devastating to the region as a whole than what happened to the New Haven Railroad. "Where yesterday fast freights and through passenger trains, including the crack Shore Line Limited, sped in rapid succession between New York and New England points carrying passengers, mail, express and the vital necessities of life (were a) of power lines, signal towers, houses, boats and thousands of tons of debris....The vital life-line between New England and points south and west had been effectually severed."[22]

Thirty-four years later, in 1972, Hurricane Agnes did $6.4 billion worth of destruction (adjusted to 1990 dollars) to that region, most of it due to flooding from heavy rains. Agnes was "only" a Category I hurricane. [23]

New England and New York today

What impact would a hurricane making landfall between New York City and Boston have on the population today? Would the ferries from Fire Island and Nantucket run fast enough to get everybody out ahead of time? Could Brooklyn be evacuated?

More than $900 billion worth of insured property—more than in the whole state of Florida—now stands in the region smashed to bits in 1938 and significantly damaged in 1972. Between 1980 and 1993, coastal investment in New York grew 216 percent; in Massachusetts, 207 percent. How much of the standing investment will be smashed next time?[24] Can anything be done to minimize the losses?

Nothing about these storms' behavior is changing. Despite gains in science and in communication, the hurricane itself has the same power, the same ability to wreck things and kill people as it did before the tremendous growth on the coast.

Camille in Mississippi

Even the most cursory account of hurricane highlights of the 20th century must include a look at Hurricane Camille, the second Category 5 of record to make itself felt in the continental United States, a $5.2 billion hurricane in 1990 dollars. Anyone still living who was in Camille's path through largely rural Louisiana and Mississippi August 17, 1969, has a tale to tell.

One of the best-known of many true, tragic Camille stories is the one about the hurricane party of twenty-five non-evacuating celebrants on the third floor of the new Richelieu Apartments in Pass Christian, Mississippi. A storm surge ranging from 10 feet at Pascagoula, Mississippi, to 21.4 feet at Pass Christian, Mississippi brought angry waves on top of the swollen high tide. One woman in the third-floor hurricane party was blown, along with her sofa, out of her window into the ocean surf and carried so far out to sea she could not see land. Ten hours later, the next incoming tide brought her back to shore five miles from where the apartment building had been. In addition to her, a ten-year-old boy in the Pass Christian party also survived, but the other twenty-three became a part of the Camille casualty list.[25]

Hugo in the Carolinas

Just at the end of two decades of fairly peaceful hurricane seasons, in 1989, twenty years after Camille, Hurricane Hugo rolled its 20-foot wall of ocean water into South Carolina. Hugo would have been far more destructive *if* he had turned south only slightly.

The "ifs," the odds and the statistics offer little comfort, however, for those who happened to be in its path. Rather than slamming into the city of Charleston, Hugo's raging eye wall slammed into a sparsely settled region north of Charleston and into a national forest. Hugo only beat up the city and doused it hard with muddy waters. He wiped out the little towns of McClellanville and Awendaw.

Of the thousands of horror stories and hundreds of personal traumas and tragedies, none is scarier than those of the hapless evacuees who left their waterfront properties and mobile homes and took shelter in Lincoln High School near McClellanville.

The storm surge charged through the schools' doors and windows. "In the band room at Lincoln High School, Jaynian White floated near the ceiling, her 1-year-old son tied to her. Someone had lifted her 5-year-old son onto a shelf. In the gymnasium, people climbed bleachers and shoved children into air conditioning ducts to escape the dark waters."[26]

Only 26 deaths in South Carolina were attributed directly to Hugo. But that is not the whole story. The hurricane hit the coast with wind gusts of 139 mph and still had winds of 90 mph when it barreled into Charlotte, North Carolina, a few hours later. It left 50,000 to 70,000 without anywhere safe to live; left 270,000 jobless; created insurance claims of $4.2 billion, making it at the time the nation's most expensive natural catastrophe.[27] The American Red Cross spent $68.6 million helping folk recover from Hugo—more than three times the amount the organization had ever spent on relief from a single disaster.[28]

Although Hugo's eye went ashore in a rural area, for six hours, the hurricane wrecked $1 billion worth of assets per hour in South Carolina and North Carolina, for a total of $6 billion in damages. Taking into account its destruction in the Leeward Islands and the Mid-Atlantic States, Hugo was a $7 billion hurricane.

Hurricane *fatalities* have been on the decline in the century, largely because of increased awareness and successful evacuations, although luck may also have played a part. Meanwhile, however, the costs of hurricane *damage*—to individuals, to businesses and to governments—have sky-rocketed to large sums.

Losses of $9 billion annually

Since 1900, hurricanes have ripped about $9 billion a year on the average out of the United States, according to Dr. William Gray of Colorado State University. Actually, ten "blockbuster" hurricanes did 64 percent of that damage, as he calculated the costs. He has warned the National Hurricane Conference and the nation through the press to expect "tremendous economic losses" from hurricanes in the future.

For any danger zone residents wondering whether their homes and other assets could end up as one of Dr. Gray's statistics, the National Hurricane Center can provide a risk report. On the basis of more than 100 years' worth of data, the analysts make statistical predictions on the frequency with which certain categories of hurricanes can be expected to come within 75 miles of a given area. On a map of the coastline, they show graphically the dangers of the danger zone.

Most of the activity overall and most of the Category 3, 4 and 5 hurricanes have occurred in the past and are projected to occur in the future in the Texas-to-North Carolina section. The Virginia-to-Maine section gets fewer hurricanes overall, and far fewer Category 3, 4 and 5 hurricanes.

Within the 75-mile distance, the odds calculated on the basis of recurring patterns (1886-1990) show the following expectations:[29]

- The southern tip of Florida can expect a Category 1 or 2 hurricane every one to five years and a Category 3, 4 or 5 every one to 15 years.
- A major section of the North Carolina coast can expect a Category 1 or 2 hurricane every one to 10 years, and the Cape Hatteras area can expect a Category 3, 4 or 5 every 16 to 30 years.

- The Florida Panhandle, Alabama, Mississippi and Louisiana can expect a Category 1 or 2 every six to 10 years and a Category 3, 4 or 5 every 16 to 30 years.

- Georgia and South Carolina can expect a Category 1 or 2 every 11 to 15 years and a Category 3, 4 or 5 every 31 to 45 years.

- Along the Texas coast, from Port Arthur to Corpus Christi, a Category 1 or 2 can be expected every six to 10 years, and one strip of that section can expect a Category 3, 4 or 5 every 16 to 30 years.

- Virginia and Connecticut can expect a Category 1 or 2 hit or brush every 21 to 30 years. Virginia can expect a Category 3, 4 or 5 every 51 to 100 years. Connecticut can expect a Category 3, 4 or 5 every 101 to 150 years.

- Maryland and Delaware can expect a Category 1 or 2 every 11 to 15 years and a Category 3, 4 or 5 every 101 to 150 years.

- New Jersey and New York can expect a Category 1 or 2 every 31 to 40 years and a Category 3, 4 or 5 every 101-150 years.

- Massachusetts' Cape Cod can expect a Category 3, 4 or 5 hit or brush once every 50 years.

- Maine can expect a Category 1 or 2 every 31 to 40 years and a Category 3, 4 or 5 every 101 to 150 years.

Significantly more people live in the danger zones now than in the past. They have more and bigger houses and more commercial operations to sustain them. Many of them have invested a high percentage of what they personally own in the hurricane-vulnerable strip along the edge of the continent. Collectively, they have made big investments in bridges and schools and other community facilities. Many thousands of them live in nursing homes. Untold numbers of coastal residents don't drive at night. The population includes many, many children, of course, most of them helpless against the worst that nature can throw at them.

Hurricanes in our nation's history have wiped out vast quantities of our physical resources, as well as thousands of people. In this book, the handful of events briefly summarized as examples of what hurricanes have wrecked in the past only hint at the damage Americans have suffered. Now that we coastal residents have put so much more in hurricanes' paths, they will wipe out even more. The damage figures are spiraling upward. Of even greater concern, however, is what future hurricanes are likely to do to humans.

Chapter 3

Living in the Danger Zone Today

Growing populations, growing risks

'I have escaped with the skin of my teeth.'

Job 19:20

On a clear day near the beaches, take a deep breath, take a walk, take a swim. A black skimmer works the incoming tide, scooping lunch. No wonder beach vacations are so popular. No wonder so many Americans spend most of a lifetime planning to move to the edge of the continent when they retire.

Tourism and retirement fuel powerful economies. Along with the vacationers and the retirees, the entrepreneurs and job seekers are drawn to the nation's shorelines, especially along the Southeast Atlantic and Gulf of Mexico.

Explosive coastal development

For the last three decades the nation's coastal regions have been in a population explosion. Demographers have worked the arithmetic in a variety of ways and devised various ways to state what has happened.

- Florida's coastal population swelled 37 percent between 1980 and 1993—from 7.7 million to 10.5 million. Now more than three-fourths of Florida's population lives in counties adjacent the shorelines of the Atlantic Ocean or the Gulf of Mexico.

- Georgia, South Carolina, North Carolina and Virginia experienced growth near their shorelines of 20 percent in those thirteen years.

- In all of the southeastern Atlantic coastal areas, in the *two decades* between 1970 and 1990, the growth was nearly 75 percent.

- At the end of the 20th century, in the regions vulnerable to hurricane dangers, populations approached 50 million.[30]

The 78 million baby boomers, just now entering their pre-retirement years, have barely begun their rush to the sea. These boomers, as well as people in their late 50s and 60s, will receive $10 trillion to $20 trillion of inherited wealth, triggering the largest transfer of assets in history. Many of them also are riding on 401(k)s and individual retirement accounts in a stock market boom. They will have options enabling them to live almost anywhere they want to live, and surveys show a majority of them want to live within 50 miles of the coast.

The technological revolution gives this group flexibility long before retirement, however, and much of the coastal population growth in the late 1990s has become what one developer called the "crest of a new boomer craze."[31]

As boomers move in with their cell phones and laptops, and as retirees arrive with their golf clubs and bicycles, economic opportunity bounces up for anyone in the construction, resort, retail or service industries. Since 1993, jobs in the 100 fastest-growing coastal counties have been created 30 percent faster than in the rest of the country.[32]

If current trends continue, the National Oceanic and Atmospheric Administration (NOAA) projects that by 2010, 73 million Americans will live in the counties fronting the Atlantic Ocean and the Gulf of Mexico.[33]

This build-up of coastal populations started during a 20-year lull in hurricane activity. Of the Texas coastal population of 4.3 million in 1990, less than one-fourth had firsthand knowledge of what happens when a full-blown hurricane hits. Similar low ratios hold for the other eighteen hurricane-vulnerable coastal states.[34] Along with having no personal experience to help them relate to hurricanes, most present-day coastal dwellers also have no familiar family history or folklore that includes hurricanes. In short, they have no realistic understanding of hurricanes and hurricane risks in the places they live.

Coastal dwellers' ignorance and cavalier attitudes frustrate professional emergency officials such as North Carolina's hurricane program planner. "People usually have to experience an event before they ever become true believers. North Carolina places a high premium on family preparedness and still people do not believe it will happen to them," he said.[35]

Decisions on whether or not to evacuate coastal populations are not about avoiding inconvenience. Mostly, they are about removing people from low-lying zones likely to be inundated by thrashing ocean water, in some places at some times peaking at more than 30 feet deep. Evacuations are called in the hopes of preventing drowning.

Indeed, as many coastal folk will say, the odds are low, actually, that a particular coastal community in any single year will be hurricane-ravaged. *That* truth, however, does not change *these* truths:

- As the planet's most dangerous weather systems, hurricanes that come ashore are always destructive and always potentially deadly.

The Saffir-Simpson Hurricane (Intensity) Scale in the appendix describes the range of power and the range of devastation they cause.

- Since 1886, on the average, every year three to four hurricanes or tropical storms have made landfall somewhere along the Atlantic or Gulf of Mexico coasts.

One example: the South Carolina Lowcountry today

Where an estimated 2,000 Americans—and maybe more—were drowned in a Category 3 hurricane in 1893, at the beginning of the 21st century, a lot of people live, work, invest and vacation. The resident population is close to one million, and growing. The annual tourist population—meaning visitors throughout the year to Savannah, Hilton Head Island, Beaufort and Charleston—is close to ten million.

Tens of thousands of houses, condominiums and hotel rooms stand on land attacked in 1893 by a 10-to12-foot storm surge on top of a high tide—a mass of ocean water topped by thrashing waves. Much of the land now developed was overrun again in the hurricane of 1940. Hundreds of thousands of square feet of retail stores and office space stand in those low places.

Considering the way the water saturated the soil in hurricanes of the past, imagine what hurricane rains and flooding will do now that so many roads, parking lots and rooftops cover so much of the territory. Consider how the water wiped out the corn and cotton fields in 1893. Then consider what it would wipe out today, if it rolled into living rooms and hotel lobbies and supermarkets and banks and real estate offices and golf courses and assisted living centers for the elderly—all now standing in what used to be dune fields and forests and swamps.

The six coastal counties in the hardest hit coastal region (Chatham County in Georgia and Beaufort, Jasper, Colleton, Charleston and Georgetown in

South Carolina) had insured properties with an estimated value of $63 billion in 1993.[36] On the basis of the 12.5 percent annual growth rate that prevailed for the five years previous to that, by 2000, the insured properties there would be valued at $143 billion. On the premise of calculations based on the booming economy and intense marketing in the 1990s, the total property values might well have reached $170 billion by 2000.

Even a huge Category 3 hurricane that came ashore on a high tide would not, of course, demolish everything in that area. And yet, property insurance companies would have to pay out a lot of bucks—nobody knows how many. In addition, other costs would hit various other victims: Business owners would be out of business for varying lengths of time, some of them forever. The region's public facilities would take a beating; the list of damages would likely include sewage and water treatment plants, electrical power systems, highways, bridges and schools. While incurring the significant expenses of cleaning up, repairing and rebuilding, state and local governments would be working from drastically reduced property and sales tax revenues.

A Category 3 hurricane making landfall again south of Savannah and charging inland in the pattern of the 1893 hurricane would wreck more than the rapidly developing coastal region. It also would create a swath of damages estimated at $32 billion, at least, across the continent from Georgia and South Carolina to West Virginia, New York and Nova Scotia.[37].

Thanks to improved forecasting and communicating about hurricanes, it is probable that most of the population would evacuate and that, with luck, the fatality list might be short. However, it is also possible that hundreds of coastal folk would stay put during an ordered evacuation, hoping that the hurricane would turn northward and not come into their community. They could be drowned or struck in the head by falling timbers. Whatever the impact on lives, there would be massive disruption, dislocation and destruction—sure to be followed by an enormous debris-disposal problem, and a rat's nest of redevelopment issues.

The pattern of growth and of risk represented by that small section is repeated again and again along the coast, from Maine all the way to Texas.

Nursing homes and hospitals

To think of hurricane-vulnerable coastal populations as primarily active, vigorous working populations, vacationers and young retirees is to think of fairyland. Nursing homes have sprung up on the coast, almost as fast as tennis courts. Among the people in harm's way, then, are many thousands of folk who get around in wheelchairs and thousands more who breathe bottled oxygen. They, like others requiring extraordinary help to protect themselves from hurricanes, are what the emergency management officials call "special needs" cases.

In Florida alone, where 78 percent of the entire population lives in coastal counties, 700 nursing homes take care of people who could not evacuate from a hurricane on their own.[38]

Add to the elderly and the nursing home residents, patients in hospitals. The folk in hospital beds, whether because of illness or surgery or injury, should not be hauled out the door on a gurney to ambulances or to helicopters to take them to new hospitals in new cities. And yet, when a hurricane landfall is imminent—or likely—what else is there to do?

If the hurricane *comes,* floods the operating room and cuts off the electricity in a hospital, patients will look in hindsight at a complex, inconvenient evacuation as an example of prudent decision-making. If the hurricane *passes by*, after a communitywide precautionary evacuation, many of these patients will remember forever the discomfort, the fright and the inconvenience they endured. Many of them will proclaim as long as they live that their evacuation ordeal was unnecessary.

Others who would be helpless against hurricanes but for adult caretakers are the coastal children, infants and toddlers as well as schoolgirls and boys. Among the numerous places of crises during a mandatory evacuation order are the coastal region's child-care centers. After a hurricane disaster, one of the supplies most in demand is diapers.

Coastal tourists

Coastal *residents* of all ages, sorts and conditions are not the only people on or near the beaches on a given day in August. Summer *vacationers* crowd the beaches of Cape May, New Jersey; Long Island, New York, and Cape Cod, Massachusetts as well as the beaches in the South.

In a report to the National Hurricane Conference in 1992, Robert C. Sheets, director at the time of the National Hurricane Center in Miami, used Ocean City, Maryland, as an example of what typically happens to beach resorts during hurricane season. In 1990, Ocean City's resident population was only 35,000, but on a single weekend in 1991, its total population was 350,000. Most of Ocean City's "guests" were on a carefree holiday.[39]

That the good old summertime and early autumn coincide with hurricane season rarely crosses the minds of vacationers. Their pleasure and relaxation are tied to property that at some point in history has been overwashed by a hurricane's storm surge and pounded by relentless winds and waves. And yet, having saved their money to rent the beach house, having squirreled away their vacation time to have a week or two at the beach, the summer tourists are in no mood to evacuate because a hurricane might come ashore where they are.

On a sunny July afternoon in 1996, the looks were grim on the faces of beach-vacationing families stuck for four hours in near-gridlock traffic trying to get off Hilton Head Island, South Carolina. As it turned out, Hurricane Bertha chose the Wilmington, North Carolina, area for landfall instead of Hilton Head. At the critical decision time, however, it appeared to be headed straight for South Carolina.

Tourism-related commerce

The businesses that serve and sell to the beach vacationers, whether on Hilton Head Island or South Padre Island, Texas, especially do not want to have people leave the beaches because of the *possibility* that a hurricane might come. For a T-shirt shop, a beach toy rental operation, even a Hyatt Regency, the difference between evacuating and not evacuating can mean the difference between profit and loss for the year. Tourism-dependent coastal commerce dreads the thought of closing up shop, sending the employees home and telling the tourists, all carrying credit cards: "Sorry, hope to see you again soon."

"Objections become more prevalent when a hurricane threat does not come to fruition," said North Carolina's hurricane planner. In those cases, storm impact is not felt but lost economic impact is. Afterward, the planner said, "Blistering editorials are written. Elections are lost."[40]

In the area of Norfolk and Virginia Beach, Virginia, the momentum of happy vacationing thousands juxtaposed against the insidious congestion on the highways almost certainly means *no* complete hurricane evacuation will occur there, according to one national hurricane official.[41] One route out is through an underwater tunnel. In the worst likely circumstances, a hurricane evacuation from that region would take at least thirty-six hours.

"Norfolk tried to evacuate for Hurricane Gloria in 1985, and luckily the storm went on up the coast. Ten miles of cars backed up in that near-disaster.

"It is frightening, I tell you. The worse place anybody could be in a hurricane would be in a car. I'm afraid that, as we are going today, we are going to get huge casualties in a string of cars somewhere sometime. There could easily be 1,000 killed," Jerry Jarrell, former director of the National Hurricane Center, said.[42]

New Orleans, Louisiana which sits four to six feet below sea level, has three exit highways, all subject to early flooding from the rise of storm tides. A pedestrian-oriented city, New Orleans has 100,000 residents without cars.[43] Easily as many as 40,000 tourists, many of whom go in and out by air, mill about the city on a given day. Under the worst likely circumstances, a complete evacuation of New Orleans would take more than fifty hours.[44]

New Orleans hotels filled with conventioneers as Hurricane Andrew barreled toward them in 1992 flatly refused to close and urge their guests and employees to get out, even when told to do so.

Fortunately for New Orleans in August of 1992, Andrew, at the time a Category 3 hurricane with winds of 112 to 131 miles per hour and a storm surge of 8 to 12 feet, swerved into eastern Louisiana sparing the city below sea level.

"Disasters: The Journal of Disaster Studies" described the city in 1996 as a disaster waiting to happen—with a metropolitan population of 1.2 million, with neither adequate evacuation plans nor safe-shelter plans in place to handle a major hurricane.

New Orleans experienced another close call in September 1998. After Hurricane Georges walloped Puerto Rico and the Florida Keys, he pushed his storm surge and his rainfall-laden storm clouds toward the Gulf coast. Tens of thousands of New Orleans residents scurried around looking for safe places. Some successfully drove out of the city onto higher ground. Some lugged duffel bags to the Louisiana Superdome and camped for a few days. Others stayed put and fretted, or partied, or prayed. But Georges came ashore and did its flooding and battering at Biloxi, Mississippi, instead of New Orleans. The city below sea level had escaped once more.

Although the Jersey coast has not been evacuated for as long as most residents can remember, the supervising planner in the New Jersey Office of Emergency Management believes that the 1.5 million population would board up and hurry inland as fast as possible if told that their lives were at risk from an approaching hurricane.[45] It would take the entire region about thirty-five hours to evacuate.[46] That it might be necessary at some point is clear to officials, who know that since 1893, twenty-one hurricanes have come within 125 miles of the coastline, for an average of one hurricane every 4.7 years.[47] The planner acknowledged the potential difficulties in convincing people to leave and added, "It is likely that any agency related to tourism would be resistant, with the casinos in Atlantic City being the toughest among this group."[48]

The tourist revenue lost from closing Mississippi's Gulf Coast casinos for an evacuation would be $5.5 million a day; from closing Atlantic City's, $13 million a day.[49]

On a given day during hurricane season, neither the casino operators nor the gambling guests are going to want to evacuate the crowded, 24-hour-a-day playing tables. The gamblers losing money are not going to want to leave. Those who are winning are not going to want to leave. The owners are likely to prefer gambling that the hurricane will go elsewhere. So who will evacuate willingly? What will happen to those who do not?

And smokestacks

Where heavy industry thrives on the coast, resistance to evacuation thrives too. Vast complexes that operate around the clock cannot shut down with the flip of a switch. Business interests dependent on continuous processes such as paper-making, petroleum refining and aluminum manufacturing do not want to be told that their employees must evacuate—based on the *possibility* that high winds, flood waters and deluges of rainfall *might* whip into town.

Shutting down and restarting the petrochemical plants along the Texas coast costs $200 to $300 million.[50]

During the Hurricane Bertha crisis in Savannah, Georgia, in 1996, the Savannah Area Manufacturers Council, a network of 349 manufacturers in the region, with 18,000 employees, groused publicly about emergency officials being "evacuation happy."[51] In early July as Bertha headed toward Georgia, she was a Category 2 hurricane that revved up to a Category 3 hurricane at one point and then mysteriously lost power and became a

Category 1 hurricane, all within a period of three days. Chatham County, which includes numerous islands east of the main city of Savannah, had an estimated evacuation-clearance time ranging from as little as 14 hours to as much as 26 hours.

On the basis of advisories coming out of the National Hurricane Center, on finely tuned computer modeling and on evacuation criteria adopted by the local governing body, at one point, county officials called for a *voluntary* evacuation of the county. Businesses, military installations, government agencies and individuals began closing down and getting out. Within the year, the county's director of emergency services was honored nationally for his judgment in this case.

Union Camp Corp., a paper and pulp industry now owned by International Paper, was the county's second largest industry and second largest employer at the time. The company immediately sent out the word in the heat of the moment that evacuation was not justified and that employees would be fired if they fled from the storm instead of showing up for work.

"Unnecessary evacuation costs a lot of money," a representative of the manufacturers' council said. "We get our own information off the Internet and each individual facility has its own detailed plan for such emergencies. Our number one priority must be protection of our infrastructure."[52]

After Bertha turned northward, bypassing Chatham County, the manufacturers' council spokesman insisted that the issue was "never lives versus money." She said the controversy between the industry and the emergency management people hinged only on "timing."[53]

Evacuation timing

Timing, of course, *is* the big problem. As in politics and dancing and sex, timing in hurricane evacuations is everything.

From the public's standpoint, at least two typical standards cause confusion, angst and argument:

1) Disaster officials adamantly recommend starting hurricane evacuations in time to finish them *before the onset of gale force (40 mph) wind.* They hear plenty of objections to this standard. Nevertheless, good reasons exist for it: Vehicles swerving and shimmying in gales can cause wrecks that block the highways. Law officers have a hard time bracing against high wind to direct evacuating traffic. Boats fleeing to safety are more likely to collide with bridge pilings once 40-mph winds start raising whitecaps in the creeks and rivers. A disabled bridge can bring a coastal evacuation to a standstill.

Loose debris flies around in gales. Imagine a family trying to load up to evacuate with lawn chairs slamming into them. Imagine a nursing home attendant trying to move a wheelchair up a ramp into a van with pine branches crashing nearby. Trees and limbs fall in gales, blocking streets, knocking out power.

2) Disaster officials also strongly prefer *starting hurricane evacuations in daylight hours in time to finish them before nightfall.* They hear expressions of outrage over this standard too. And yet, good reasons exist for it also: It is almost impossible to get a message to the masses after they have gone to bed. Late at night, instead of during the day, it is harder for those who should evacuate to find the hurricane survival gear, board up the windows, stow the outdoor furniture, make arrangements for the dog, pack the vehicles and get on the road.

In emergencies, people are tenser in the dark than in the daylight—and more prone to make mistakes. People function less effectively at the end of the day than at the beginning. Wrecks—that could bring critical evacuations to standstills—are more likely to occur at night.

Even in the face of a serious hurricane threat, most people do not start evacuating after nightfall. Most take the position that they need a good night's rest to prepare them to face whatever ordeal they might have to face the next day. Most go to bed, expecting to get the latest hurricane information early next morning and to decide then whether or not to leave the coast. Their fondest hope, of course, is that the morning's reports will show the hurricane headed somewhere other than where they live.

Realistically, decision-makers have to take into account that coastal populations simply do not evacuate in mass between midnight and day-light.[54] For that reason, among others, evacuation decisions sometimes have to be made many hours in advance of a potential hurricane landfall.

As an illustration of the timing problem, consider this hypothetical scenario: Suppose a coastal community could be thoroughly evacuated in 12 hours. Suppose forecasters could accurately predict the hour of the landfall of the eye of the hurricane 18 hours in advance. Suppose forecasters also could predict with confidence that the gale force winds on the outer edge of the hurricane would blow into the community in 12 hours.

If the forecasters' prediction of a direct hit at midnight could be made at 6 a.m., the community could be emptied of people during the daylight hours between 6 a.m. and 6 p.m. The evacuation would be concluded just as gale force winds arrived, about six hours before the eye came ashore. By the time the storm surge hit land, everyone would be safely tucked away on high ground.

Now suppose the forecasters' prediction was made at 6 p.m. instead of 6 a.m. The people would need to start leaving immediately. If it would take 12 hours for everybody to get out, the danger zone could not be emptied until 6 a.m., at best. Nine hours of that evacuation would have to occur in the dark. That would be the theory, but it probably would not occur, simply because most people will not evacuate at night. If the evacuation were postponed until daylight the next day, the community certainly would be overrun with a storm surge before everybody could get to safety.

Even with the hypothetical advantages of a 12-hour evacuation time and forecasting skills truly accurate 18 hours ahead of the hurricane, coastal populations still would need to get used to the idea of leaving when the sun is shining and when the hurricane might not come. For many a coastal dweller, the only sensible thing to do during a hurricane threat is to start evacuating early—during a period of many uncertainties.

Considering the dilemmas that would occur even under the best imaginable circumstances, it is easy to see the difficulties in the real world. In the real world, few if any, coastal communities can be emptied of all people within 12 hours. In the real world, 12 hours in advance of the hurricane, the forecasters can say only that the hurricane is going to strike somewhere within a *100-mile stretch of coastline.* In the real world, 24 hours ahead of a hurricane, the forecasters can say only that the hurricane is going to strike somewhere within a *200-mile stretch of coastline.*[55]

Some communities should clear the coast when a hurricane is headed ashore, but which communities? Trying to clear great swaths of coastline all at once can create traffic disasters even when the hurricane goes elsewhere.

Hurricanes differ in size, intensity and speed of movement. Communities differ in population, topography and highway capacity. As the hurricane and the communities begin to intersect with one another, the dynamics of decision-making get far more complicated than most coastal folk realize.

The Hurricane Opal lesson

No community would choose to be an example of the terrible possibilities linked to the timing problem, but the experience of the Pensacola, Florida, area with Opal in 1995 serves well to demonstrate them.

As the Saffir-Simpson Hurricane Intensity Scale shows (see appendix), the more intense the storm, the higher the "category," the faster the winds and bigger the storm surge, the more people and property endangered.

In the wee hours of October 4, Opal played the dirty trick of increasing her wind velocity. In addition, Opal picked up forward speed and shifted her aim toward Pensacola. Watching a fast-moving Category 4 hurricane, instead of a slow-moving Category 3 in the Gulf of Mexico, the National Hurricane Center had issued a "hurricane warning," (expect hurricane conditions within 24 hours) at 10 p.m. October 3, anticipating evacuation orders to be called by daylight.

Next morning at dawn, as soon as people turned on their radios and TVs and consulted the Internet, they began jamming the highways. The official evacuation time for Pensacola area residents was estimated at 19 hours. In no way could everybody get out of the danger zone before the onset of gale winds and before dark.

The "dirty weather" gathered. As the winds picked up their pace and the waters began to lap at the causeways, the roads and highways filled with the cars and trucks and minivans of 100,000 people trying to get to high ground. Several construction projects on crucial Interstate 10 turned what would otherwise have been a two-lane exodus into a one-lane gridlock. By noon that day, every road leading from the beach and all the cloverleaf intersections onto the interstate were clogged.

In the frantic hours of October 4, the fleeing evacuees from one county ran straight into the fleeing evacuees from another. Traffic came to a near halt for many miles. Frustration in the stalled traffic led some evacuees to jump out of their cars and walk to houses along the road asking for shelter.[56]

The eye of the hurricane made landfall in the late afternoon at Pensacola Beach. By then, it had weakened to 115 miles an hour. Although it hit at low tide, its storm surge plus 10-foot waves threw 5 to 21 feet of water along almost 130 miles of shoreline.

Miraculously, only one person in Florida was killed, although Opal's Florida damage alone included destruction of 3,342 homes, wreckage of 14,411 homes and the dreary loss of power for 500,000 residents. Before Opal had finished with the Southeast, she had caused two more deaths in Alabama, five in Georgia and one in North Carolina. These were in addition to the earlier 19 deaths in Mexico and the 31 in Guatemala.

Although Opal struck the Florida Panhandle on low tide and diminished in power before making landfall, coastal Floridians were horrified when they returned to the beaches and saw the evidence of the power of wind and waves that had nearly attacked them in their homes.

For the 350,000 people living in the Pensacola area at the time, Category 3 Opal experience came close to being a catastrophe complete with hundreds of fatalities. For the hurricane forecasters, Opal was "the hurricane from hell." Out of what happened *that* time, the Florida Division of Emergency Management organized a new communication network in the fervent hope of improving the timing and the coordination *next* time.[57]

Problem is that next time the situation will be different: The Pensacola people might be ready for another Opal. However, the same kind of hurricane will not attack in the same way at the same time of day under the same conditions. One thing for sure about the next time is that the population will be greater and the evacuation clearance time longer.

On a clear day, Americans can see forever along the coastlines.

On a hurricane day, the scene is a jumble of elements battling one another, politics, economics, fright and complacency, along with the lives of coastal dwellers suddenly turned upside down.

Chapter 4

As 'Dirty Weather' Gathers Today

Hurricane evacuations become major, costly ordeals

> *"Decision is a sharp knife that cuts clean and straight;*
> *indecision, a dull one that hacks and tears*
> *and leaves ragged edges behind it,"*

Gordon Graham

Coastal hurricane evacuations are relatively new phenomena, a bit like bicycle helmets and seat belts. When coastal populations were concentrated away from the beaches and on high ground, when coastal populations were sparse, when the dreaded stormwater did not have to flow over square mile after square mile of pavement before reaching the sea, hurricanes did less damage than they do today.

Today, in part because of the size of coastal populations, when a hurricane approaches the nation's Atlantic and Gulf coasts, anxiety rises and urgency builds as never before. Forecasters have significant challenges. Traffic engineers have nightmares. The media have monumental responsibilities. Disaster relief organizations must gear up. And the folk with one of the heaviest loads of all are those in each region who must decide whether or not to recommend or mandate evacuation.

Will that hurricane come ashore? Where? When? How bad will it get? If people need to leave the coast to keep from being drowned in the storm surge, the questions then are: Which residents, who are they, where are they located, how many of them are there? How smart, how informed and how courageous in a potential firestorm of controversy are those who decide whether or not to evacuate whole cities? How many hours between the time the decision is made and the last person can be out of the danger zone? Is it riskier to cause massive traffic jams and grumbling louder than thunder and then see the hurricane go elsewhere, or to let people continue to take care of business or swing in the hammocks and play golf until all the facts are known? In many instances, the relevant facts cannot be known until it is too late.

The traditional rough estimate of the cost of evacuations is $1 million for every mile of coastline, but that figure may be too low. Millions of dollars are involved in silencing big-ticket operations such as paper mills and casinos, and everybody affected by an evacuation bears some cost. For a few days nothing is normal. A hotel telling its guests to go home and a beach town t-shirt shop closing its doors are kissing revenue good-bye. Some evacuees need ready cash or solid credit to take care of themselves in motels inland—for unknown lengths of time. Others depend on public shelters, which also cost money to run. Government agencies have to switch from routine to emergency mode, meaning overtime shifts at premium pay for employees. Whether the hurricane comes or not, during a hurricane warning, relief organizations swing into action to be ready, moving equipment and supplies to distribution points—just in case.

No one wants to evacuate unnecessarily. And yet, the dilemma that will not go away is this: By the time the forecasters know for sure where the hurricane is going, in many regions, too little time is left for everybody to get to safety before the hurricane hits.

Keep in mind two crucial standards: 1) Evacuations are best executed in daylight hours; 2) Evacuations need to be completed before the gale-force winds arrive.

As evacuation becomes a strong possibility, National Hurricane Center forecasters wrestle with their end of the problem—the science. Threatened residents wrestle with theirs, buying batteries, filling gas tanks and tuning in to radio and TV, hoping that if somebody has to evacuate it won't be them. Typically, as the probabilities of a direct strike go up, local or state officials may recommend that the tourists leave the beaches and that coastal residents who might be planning a trip inland anyway take that trip immediately. By reducing the population at risk, they reduce the amount of time it takes to evacuate and give themselves breathing room for hard decision making.

At some point, the call would have to be made for a "go" or a "no go." It comes down to three things: What are the odds that the hurricane will roar into a certain region within a certain number of hours? How long will it take for the endangered population to flee from the sections of the coast most likely to take a hit? Where else can flooding and high winds be expected?

Once those questions are answered, it comes down to two more things: How much risk is too much risk? What is the decision of least regret?

For thousands of thinking coastal residents, warnings and evacuations trigger pressing demands, all essential, none optional, few simple. Those who have their lists, plans and priorities figured out ahead of time can manage if they have the time and energy. Those who have not thought out their hurricane strategy can find themselves in a panic—and find their own lives as well as their property in danger.

Under mandatory evacuation, turmoil erupts in every household, every institution and every business in the communities affected. What is going to happen to the endangered beach vacationers who have no flight reservations until next week and have no way of leaving the coast? The around-the-clock industries have to shut down. Airplanes need to be secured or relocated. Before people drive away, many of them have to board up their property.

In the few hours before joining the evacuation traffic, boaters should be moving their craft into sheltered coves, battening down hatches, lashing sails, doubling anchor lines. Once the winds begin to shriek in the riggings, it's too late to secure the floating docks. Once boats and wharves and waterfront structures begin cracking up in the storm, their parts become battering rams that help the fast-moving, thrashing water beat up everything else.

When will the drawbridges be locked down, and how will the owners of the tall-masted yachts get the word?

When will the public shelters be opened? Who will make sure that food is on hand, that the power backup systems work and that light blankets are stacked in convenient places for troubled hurricane evacuees?

Is the local emergency operations center located in a secure building, one that will not flood, not leak and not lose its power? This is to be command central, the communication hub for tens of thousands of stressed and distressed hurricane survivors. It needs to function. It must not flood. It needs to hang onto its roof. Tired, sleepy emergency personnel should not have to deal with flying glass from shattered windows.

Free the school buses

Early on, the schools on the islands and in the flood zones must be closed. School buses are needed to transport military personnel—and others. In a full-blown evacuation of Beaufort County, South Carolina, school buses are packed with seven thousand U.S. Marines from low-lying Parris Island, a place nearly obliterated by the August 27, 1893, hurricane. In other areas, including below-sea-level New Orleans, where 100,000 residents do not own cars, the plan is to move as many people as possible to safety by public transit, school buses and the National Guard convoy.[58]

In addition to freeing up the buses for transportation, closing schools frees up teachers to go home to put on their window coverings, pack their photo albums and start traveling out of the danger zone. It also frees up students—to go home to get ready for the hurricane. But when will Mom or Dad be home to greet them? Dad could be a policeman with traffic-control assignments in the evacuation. Mom could be a banker working extra hours to make sure people can get the cash they'll need to leave town. Who will be available to look after the frightened children? What is going to happen in the day care centers?

Schools inland, often designated as public shelters, also must be closed, their students sent home. The American Red Cross says it can open a shelter in four hours after being notified. That means somebody must be responsible for the keys to the buildings. Someone else needs to post signs along the evacuation route telling where these shelters are on the back streets of little towns and inland cities and rural neighborhoods. Someone should make sure the power backup systems work, the bathrooms are stocked with toilet tissue, the cafeterias furnished with cheese, bread, canned beans, canned peaches and paper plates.

Historically, most hurricane evacuees traipse inland to the homes of friends or relatives or to motels, if they can find vacancies. Only ten to twenty-five percent seek refuge in public shelters. But even those numbers can be large—and they swell suddenly and uncontrollably. During the Hurricane Fran evacuation in North Carolina in 1996, as one example, 29,000 evacuees poured into Red Cross shelters within only a few hours.

Florida hurricane officials know that most evacuating residents will try to find motel rooms, but they nevertheless estimate they have a statewide shortage of more than 300,000 public shelter spaces. Although they hope never to need all the shelters at once, Hurricane Georges in 1998 forced evacuations in the Keys, sections of the Miami area and the Panhandle all within a couple of days. Hurricane Floyd forced evacuation of the entire east coast of Florida in 1999. With more public shelters, the Floridians would not have had to drive as far as they did—nor would they have had to spend as much time on the road.

After the Floyd evacuation, Gov. Jeb Bush asked the question: "Has population growth gone faster than our ability to build roads? We started behind the curve."[59]

An intense storm that crossed the tip of the Florida peninsula, either from the Atlantic to the Gulf of Mexico or in the other direction, would be dangerous to both coasts. Even the tourist-oriented Orlando region, with its hundreds of thousands of motel rooms, could not accommodate everybody trying to flee from south Florida at once.

Taking into account specific state surveys showing that 20 to 25 percent of the population will flock to public shelters instead of motels and private homes, Florida is looking at a shortfall of 154,372 spaces in fast-growing southwest Florida alone.[60]

Bold recommendations to address the shelter deficit came forward in Florida after Hurricane Andrew in 1992. New schools and college buildings, a commission said, should be *built* to withstand hurricanes, and existing schools and colleges should be *retrofitted* to make them safe hurricane shelters. Except in rare instances, neither has happened.[61]

Open the "host" towns

As a hurricane moves toward shore, in addition to the Red Cross staff and volunteers opening the inland shelters, the residents, the police, the motel and restaurant owners and the emergency room staffs of the towns 100 to 200 or more miles inland need to know the coastal evacuees might be on the way. Suppose the traffic backs up at some little-known, rural crossroads that now must be used by long lines of vehicles filled with people trying to get away from a hurricane. How will the jam be cleared? Suppose these refugee families get hungry, sleepy, thirsty. Suppose some of them suffer injury or illness or heart failures because of sheer panic. How will the south Georgians manage when the Florida Panhandlers crowd into their communities? What happens in Atlanta when the lodging is filled because of the evacuation of three states' coastal towns?

During any coastal evacuation, from all over the state, highway patrol officers must head toward the hurricane to direct traffic, to assist motorists whose cars overheat, to handle accidents on the clogged highways, to wave their arms at the key intersections as thousands of tense drivers pass through. Some drivers will stop to argue with whoever is motioning them to go on. Some will have genuine emergencies that the officers must try to address: childbirth, forgotten medications, a flat tire, anger and fright, to name a few.

Without fail, during an evacuation, motorists want to know why officials don't "open the other lanes." As they sit or creep in long lines of traffic *leaving* the beaches, they look longingly at empty pavement leading *to* the beaches. Afterward, they demand to know why somebody didn't do something to speed up the process.

The most common rationale for not opening the second lanes of two-lane roads and the third lanes of four-lane roads has been this: Emergency vehicles must have access in the opposite direction. Opening and closing of the "reverse lanes" would create massive problems in communication, in traffic control and in highway marking. Tired drivers can become frustratingly disoriented when they cannot read the signs on the side of an unfamiliar highway. In a lot of regions, reversing traffic lanes would simply move the traffic tie-ups farther up the road as traffic was forced back into its normal flow farther inland. Making such a significant change in traffic rules could create numerous and potentially deadly hazards for drivers already under duress.

South Carolina Gov. Jim Hodges heard all of these arguments before the Hurricane Floyd evacuation of September 14, 1999. Then he flew in a helicopter over I-95 and I-26 and looked down on near-gridlock as 800,000 South Carolinians tried to travel inland from the cities of Hilton Head Island, Charleston and Myrtle Beach and everywhere in between. He immediately ordered opening the eastbound lanes of I-26 for westbound traffic in the hopes of relieving the tension and possibly saving lives as well. That move eventually cleared the immense traffic jam, but not for several hours after the decision was made.

Although experienced transportation officials in every state have resisted the idea of reversing lanes for evacuation traffic, the Hurricane Floyd experience wiped out most of that resistance. After looking hard at the data on population, highway capacities, evacuation clearance times and well documented margins or error in hurricane forecasting, they are figuring out *how to manage* reversals in the future instead of explaining reasons for not doing them. Reluctantly but realistically, they consider ways to store, handle and position thousands of orange highway-marker cones and ways to get word to the public that eastbound lanes are going to carry westbound traffic. All of them hope against hope that they will not have to take those steps, but many of them are preparing to do it nevertheless.

Livestock, dogs and cats

Pets should be taken with evacuees, but most public shelters and most motels do not allow pets—and for good reason. Horses and other livestock, zoo animals and humane shelter dogs and cats that live in the storm-surge zones might be relocated—but probably only if arrangements have been made ahead of time. Sometimes inland kennel and stable arrangements can be made on the spot, but sometimes they cannot. Comes the dilemma of whether to leave Buddy the terrier in the house with food and water and a cabinet to climb onto in case of flooding. Roaring winds and rising waters, in the absence of human comfort, can agitate cats and dogs and cause them to hurt themselves trying to escape. For pet lovers, anxiety and guilt can intolerably exacerbate stress over the dangers their pets will encounter if caught in a hurricane.

After the Great Sea Island Storm in South Carolina, one hurricane survivor wrote a letter listing his assets that had been swept away—his barn, most of his house, all of his livestock. But the saddest part, he said, was the loss of his two dogs, Dan and Sheba.

Business data

For businesses during an evacuation, someone must look after the computer backups and the active records, if nothing else. With only two hours to remove essentials before the onset of Hurricane Hugo in Georgetown County, South Carolina, in 1989, attorney Jimmy Chandler got only the disks and one server and monitor from his small office in a grade-level waterfront building. When he returned, virtually everything was ruined—copier, books, files of old cases, furniture, building. "And everything was salty. I thought we'd never get through cleaning out rusty paper clips," he said. As a specialist in environmental law, Jimmy Chandler learned a new lesson in environmental awareness that week.[62]

In every office, as a minimum, someone should look after the customer list and the accounts receivable. It's no good getting back after an evacuation, hoping to crank up business again and finding basic business records—without which business is nigh impossible—gone or wet and scattered.

The banks need a plan for hurricane evacuations, as do the insurance companies and the appliance repair shops. Newspapers, expected to publish special hurricane editions no matter what happens, need plans for their news staffs, their press crews, their circulation team, their expensive, critical equipment.

The stories of what happens to small businesses in hurricane disasters are ghastly. One survey found that only 50 percent of the small businesses located in a community slammed hard by a hurricane are able to recover successfully from the blow. In some cases, hurricane plans developed in advance make the difference between demise and survival.

What happens in a hospital

Of all the organizations challenged by hurricane emergencies, few have as many life-and-death calls to make as hospitals and nursing homes, especially those in the storm-surge evacuation zones. Long before every hurricane season, agreements should be in place for the relocation of patients to inland facilities.

Three examples illustrate the resourcefulness required:

- The roaring winds of Hurricane Andrew outside Deering Hospital in Miami in August 1992 overpowered the voices of the doctors, nurses and patients. While the hurricane pounded, the staff moved the sick, the injured and recent surgery patients to central areas of the building, carrying intravenous equipment and tubes and medications and charts, giving up pieces of the hospital compound—rooms, corridors, whole wings—as they relocated. By the time the storm subsided, Deering's emergency room was wet and abandoned. Frightened emergency room staff and patients had relocated too, while Andrew had pried loose parts of the building and poured rain into it. Deering closed for repairs after the storm and did not reopen until nine months later.[63]

- Medical University of South Carolina's teaching hospital in Charleston lost windows during Hurricane Hugo, forcing nurses to huddle patients into corridors while orderlies leaned against the doors to keep the wind and rain away from the beds. The hospital's chief said the room he was in felt as if "it was breathing," so strong were the pressure fluctuations while the hurricane winds raged through the city.[64]

- A Jacksonville, Florida, hospital sits on a site expected to experience a 17-foot storm surge in a Category 4 hurricane. That staff there plans to evacuate patients upward to the second floor when a hurricane hits Jacksonville.[65]

To keep from compounding the natural disaster with a hospital disaster, serious thinking, planning and rehearsing have to be done in advance. Successful evacuation of the patients of hospitals in evacuation zones to other hospitals depends on pre-negotiated mutual aid agreements.

Things happen so fast (once the storm's fury is imminent, then raging just outside the window) that without plans worked out and drilled into staff calmly ahead of time, not even the quickest thinking professionals can make sense out of what a hurricane does to a hospital or nursing home. It is fine to have a power generator, for example, but the generator is no good when the fuel runs out and no good with a broken part for which there is no replacement. It is fine to have the second floor of a hospital intact so that patients will not get wet when a storm surge rushes in, but it is no good to be unable to medicate and feed them—either because the kitchen and pharmacy are on a flooded first floor or because supplies are unavailable.

Fleeing to what kind of shelter?

Widespread has been the assumption that school buildings work as emergency shelters. In truth, many schools in coastal communities are subject to flooding or even storm surge, and few anywhere are designed and constructed to resist hurricane-strength winds. Those with flat, lightweight roofs and wide eaves are especially vulnerable to being decapitated in a storm.

"Any shelter in a storm" is not operative. Buildings that flood or fail in other ways endanger their occupants.

When disaster specialist Pat Goodale of the American Red Cross reached McClellanville, South Carolina, after Hugo in 1989 and found the condition of the Red Cross shelter there she realized that the nation's chief disaster relief agency had unintentionally "put people in harm's way." Ocean water had charged across Bull's Bay into Lincoln High School's gym. People there had assumed they would be safe, but they were not. The 20-foot storm surge forced evacuees to scramble into the rafters. The frightened evacuees who nearly lost their lives in a Red Cross shelter include elderly men and women and young children.[66]

The Lincoln High incident forced the Red Cross—the nation's official disaster relief agency, according to Congress—to reassess its shelter-location guidelines. "We cannot in good conscience encourage people to leave their homes and go to places that are not safe," Michael Logan, national Red Cross hurricane coordinator, said.[67]

Soon after Hugo, Red Cross officials at national headquarters reduced its list of approved shelters, kicking off those subject to flooding in a Category 4 hurricane. They also established for the first time wind-resistant standards for buildings that the Red Cross would operate in hurricane emergencies.

In addition, the Red Cross officials at national headquarters decided that buildings that would not be flooded but that *would be isolated because of flooding elsewhere* also could not be used as Red Cross shelters.

The Red Cross does not want to take responsibility for evacuees in any shelter that it cannot "service" after the storm, Pat Goodale explained. "If the place is flooded, or if the roads leading to it are flooded, we can't get food, supplies, medical necessities and emergency equipment into it. Suppose a shelter catches on fire. We have to think about what is going to happen after the hurricane is gone."

The new rules about shelters and the shorter list of qualified places for sheltering hurricane evacuees created a firestorm among the emergency professionals who understood the significance of what had happened—in light of the growing coastal populations.

Robert C. Sheets, former director of the National Hurricane Center, for one, argued at the time that it is better for people to save their lives, even if their feet get wet, than to die in a hurricane.

In some places under some conditions, for some people, going upstairs in a hospital or a courthouse or a hotel or a school building may be the only possible sanctuary. While it may be unnerving to become a fugitive on the second floor of an unfamiliar building as the ground floor fills with water, in certain circumstances, it might be the one safe, available alternative to trying to swim in the storm surge. For the Florida Keys, New Orleans, Norfolk-Virginia Beach and New York City, some emergency management experts are thinking about what they call "vertical evacuations." Even less secure than the upper stories of flooded buildings are what they call "shelters of last resort," buildings not to be identified publicly ahead of a hurricane but to be used only in the worst possible circumstances. Both vertical evacuations and shelters of last resort are fraught with problems, both the kind that are obvious immediately and the kind that will not come to light until they are used. However, in some places, no better options wait in the wings. When there is no time left to flee to high ground, what else is there to do but to try scramble away from a hurricane?

About the "least-risk" guidelines the American Red Cross established for the shelters it would sponsor and take care of, Dr. Sheets said, "They are not being realistic." In his mind, population growth has so outpaced forecast skill improvements that the relief agencies must get used to the idea that many shelters are less than ideal. They should plan, he said, to provide relief services to them as best they can.[68]

Charles Chesnutt, the U.S. Army Corps of Engineers' coastal engineer in charge of hurricane evacuation studies, describes the change in thinking under way. In New Orleans, for example, he says, "We have known for a while that we cannot get 500,000 people out of there. It would take 60 hours, and nobody is going to start evacuating 60 hours ahead of a hurricane because there are too many uncertainties about where it will come ashore.

"So now the question is this: 'How many people can get out, say, in 24 hours, and what are the rest going to do instead of leaving the city?' "[69]

Reluctantly, because of the life-threatening nature of the dilemmas in New Orleans, Red Cross hurricane planner Michael Logan has discussed with local players in emergency management the possibilities of vertical evacuations in that city. He said he could imagine a group of frightened evacuees on the second floor of a building with the water 17 feet deep with snakes and alligators swimming around in it on the first floor. It is a scene he hopes neither the Red Cross nor the people of New Orleans will face.

When the danger bears down

At the beginning of the 21st century, the direct aim toward the continent of a hurricane triggers a barrage of activity. Hurricane Hunters aircraft, stationed at Biloxi, Mississippi, fly in and out of the hurricane twice a day, at 10,000 feet, radioing their data to meteorological analyzers. The National Hurricane Center's reconnaissance jet, Gulfstream 4, drops sophisticated data-gathering devices called dropsondes from 45,000 feet up. Satellites and radar transmit images and data continuously to meteorologists and to the public.

The Weather Channel tells and retells whatever its reporters can find out and show picture after picture of pounding surf and blowing palmetto and palm trees. Various web sites put out data, show satellite and radar images, publish advisories and post updates constantly. The nation's certified "hurricane nuts" and certain coastal industries work their own hurricane-tracking computer software to make private predictions about where the hurricane is going and how big and bad it will be if and when it hits land.

Adrenaline starts to flow in the multitudes of emergency preparedness officials drawn together previously by obligatory hurricane-planning ventures. Calls are heard for watches, then warnings, then evacuations. Various professionals, including law enforcement officers, disaster-management specialists and elected officials, convene in Emergency Operations Centers (EOCs), then regroup and regroup, responding to every nuance of hurricane movement.

In the coastal region predicted as the most likely target of the hurricane's landfall, traffic jams highways and bridges. Discount stores run out of batteries. Building supply stores run out of plywood. Many residents check the "premium due" date on their flood insurance policies, pack their photo albums and telephone relatives and motels inland for spend-the-night reservations. About 20 percent head toward public shelters.

Some might not get out in time, simply because of the crush of traffic.

While many are leaving or trying to leave, others elect to stay put, hoping the hurricane will swerve, anticipating that this hurricane, if it hit them, will be exciting but, with a little bit of luck—not too harsh. Surely those who choose to stay put are only those who never heard about what happened to South Carolina in 1893, or to Texas in 1900, or to the Florida Keys in 1935, or to Mississippi in 1969.

Long ago, ships' captains steered away from dangerous storms when they could figure out where they were and guess accurately where the storms were going. Long ago, coastal settlers mostly built their homes away from the beaches, on the mainland instead of on the islands, back away from the waterfronts.

Today, ships' captains use a vast array of science and technology to avoid hurricanes' paths. Today, coastal settlers pay big bucks for property known to have been lashed hard by hurricanes in the past and sure to be lashed again.

For those living on the very edge of the continent now, all might go well for a long time, even a lifetime. Comfort, however, turns to anxiety and to urgency and then to emergency and then to a crescendo of mass frenzy when a coastal region finds itself in the target zone.

Somewhere, at some time, some coastal residents and perhaps some coastal vacationers as well are going to face up-front and personal a mountain of ocean water driven by hurricane-force winds. They could drown. Whose tragedy it will be depends on who will not—or who cannot—evacuate in time.

In the hopes of avoiding such a tragedy, the nation's hurricane scientists ache with the burden of producing accurate, timely information for the public. They are getting better and better at saying what a hurricane will do within the next few hours. What they most want the vulnerable coastal populations to understand, though, is that the foreseeable future shows continuing forecasting uncertainties further than a few hours out. The further out evacuation decisions must be made, the more uncertainties the evacuation decision-makers face.

The more sharply the danger zone residents see the forecasters' problem, the better their chances for making sound decisions and saving themselves from unnecessary grief.

Chapter 5

Forecasting Science under Pressure

Partly cloudy in modern times

> *Why didst thou promise such a beauteous day,*
> *And make me travel forth without my cloak*
> *To let base clouds o-er take me in my way.*

William Shakespeare, Sonnet 34

Despite rapid advances in meteorology and technology since the 1900s, hurricane forecasters in the 21st century work on a more exposed edge than ever before.

When a hurricane aims at land, all eyes turn toward the National Weather Service's National Hurricane Center in Miami. Suddenly meteorology becomes a stressful science. The safety of the public depends to a large extent on local and state officials, who are in turn dependent on the forecasters.

Hurricane science has moved steadily forward for a century, and yet, as far in advance as the information is needed for absolute safety, the forecasters cannot *guarantee* a projected track. Nor can they be *sure* of ground-level wind speeds once the hurricane hits land. They know a lot about the storm surges, but they *do not know precisely* where they will occur because they *do not know exactly* which track the hurricane will take. They can get a lot of data about moisture in the clouds, but they *cannot tell for sure* which yards will be flooded, which roads will be washed out because of heavy downpours. How fast the hurricane will be traveling when it strikes the coast is one more variable, always subject to change.

Despite giving these phenomena more attention than any other group of scientists in the world, the National Hurricane Center's forecasters cannot point to a dot on a map and swear by all that is holy that a hurricane, at such and such a strength and size and speed, is going *there at a time certain*.

'Dirty hands' science

Jerry Jarrell, former director of the National Weather Service's National Hurricane Center, described the hot seats on which the forecasters sit:

"My guys do 'dirty hands' work. We wrestle with the 'disconnect' between the research and the applications. 'Real' scientists would go crazy under the pressure. This is still an inexact science, but we have to make decisions based on the best information we can get."[70]

The best information the meteorologists can get nowadays comes from an array of devices in various locations around the planet. Since the era not so long ago in which forecasters depended on ships' crew members to let them know in person about offshore hurricanes, the science and the technology have come a long way.

In the very early 1900s, wireless telegraphs began transmitting ships' messages to what was then called the Weather Bureau, a division of the federal Department of Agriculture.[71]

By the 1930s, a network of data-collecting, air-borne balloons had evolved.[72]

In 1943, an Air Force pilot and navigator made the first recorded pre-meditated flight into the eye of a hurricane. A wild ride, no doubt, their trip began the Air Force's aircraft reconnaissance operations, today a vital part of the hurricane research and forecasting. Nowadays, from Biloxi, Mississippi, the famous "Hurricane Hunters" fly regular 5,000-to-10,000-feet-high missions through hurricanes. They beam the data taken at 30-second intervals to satellites, which beam them to ground-level computers.[73]

The coastal radar network has been evolving since the mid-1950s. Today, radar provides nearly continuous coverage of storm activity within about 150 miles of the shore from Maine to Texas.[74]

The more refined Doppler radar network put in place in 1990 can measure the amount of water in a hurricane's clouds and give close estimates of wind speeds at various heights except ground level. Doppler has spotted the eye of a hurricane 200 miles away. It has enabled the scientists to make extremely accurate forecasts two hours ahead of severe conditions. Although two hours is no good for hurricane evacuation purposes, it is invaluable for warnings against the flash floods from rains and against tornadoes that accompany hurricanes.[75]

Satellites take the pictures

Since 1975, the GOES (Geostationary Operational Environmental Satellite) system has been producing images that span the entire hemisphere at 30-minute intervals. Today it is the backbone of the Weather Service's day-to-day data gathering. The polar-orbiting satellites at lower altitudes pick up supplemental data. These satellites beam down infrared photographs that make terrific images for the lay public's viewing. More importantly, they give the meteorologists the "big picture" as nothing else can.[76]

Prudent captains of ocean-going vessels avoid storms these days, but the data *they* can furnish the hurricane forecasters still is welcomed and is used nevertheless. Today, however, the bulk of surface data about conditions in the Atlantic Ocean and the Gulf of Mexico—temperature of air and water, atmospheric pressure and wind speed—comes from anchored oceanic buoys.[77]

The meteorologists pin a lot of hopes for data they say they have needed for a long time on a reconnaissance jet, the Gulfstream 4, that began flying around and into hurricanes in 1997. Two pilots, two observers and eight scientists are assigned to it. Flying up to nine hours, as high as 43,000 feet, the jet drops 13-ounce instruments called dropsondes. They continually measure air pressure, temperature and humidity as they fall to the surface. Like the lower-level "Hurricane Hunter" aircraft, the jet's dropsondes transmit the readings to satellites, which beam them to satellite receivers on the ground and thus to computers for analysis.[78]

The new ability to take the vital signs of the hurricane, at varying levels, and the vital signs of the air just outside of the hurricane, also at varying levels, the meteorologists believe will help them refine their predictions.

The data-gathering equipment feeds continually into a super computer in the National Center for Environmental Prediction in Maryland, where basic computations and analyses are done. Sophisticated computer models take the analyzed data from around the world to create a range of predictions. The varying predictions appear on the monitors of the forecasters at the National Hurricane Center on the campus of Florida International University in Miami.

There, eight software programs that track hurricanes are in use, each different from the others. Often, the tracks they predict from the same data and same analysis process are similar but not the same. On occasion, the projected tracks vary all over the map.

National Hurricane Center

Forecasters provide information and advisories on the basis of what they see on their computer monitors and how they interpret what they see. For bursts of time, they scurry from model to model, comparing, contrasting, drinking a lot of coffee and responding to a barrage of questions and advice. They bring several layers of expertise into their discussions, knowing that much is at stake, knowing also that no matter what they say and do, they will be second-guessed.

Some, but not all, tropical depressions become tropical storms, and some, but not all, tropical storms become hurricanes. Some of the systems travel steadily westward or northward. Some meander here and meander there. They tease, and surprise, and sometimes even startle, the meteorologists who know the most about them.

Which atmospheric elements are controlling the hurricane's movement, and where are they taking it?

How big is it? How far out do its hurricane-force winds swirl? How about the gale-force winds?

How intense is it? What are the wind speeds in its various sections? Are they changing?

How fast is it traveling?

How big is the eye, and therefore how big is the eyewall, the swirling band of the hurricane's strongest winds?

Track: Where will it go?

One of the eight tracking models in use, the "no-skill" (CLIPER, CLImatology and PERsistence) forecast model, fed with appropriate historical and climatological data, develops *statistically* sound track projections. Until the 1980s, it was the most reliable tool available for predicting a hurricane's path. For the public trying to understand what is going on, it has the advantage of relative simplicity.

Unfortunately, almost nothing about hurricane forecasting is actually simple. Along with CLIPER, the forecasters now also use more complex, numerical models, some of them taking into account much more data. Altogether, the newer forecasting models, the increased volume of data available and the development of the interpretive skills among the scientists have improved track predictions 26 percent in the last 26 years. The meteorologists are proud of that achievement.

They do not expect that element of hurricane forecasting—the path—to get much more precise soon, however.[79]

"We had the best forecasts ever in 1996. It is going to be difficult to improve on that," forecaster Miles Lawrence said.

In that "best" year for forecasts that Lawrence touts, Georgians and South Carolinians evacuated their coastlines twice—for Hurricanes Bertha and Fran. Both hurricanes toyed with Georgia and South Carolina but slammed into North Carolina. Does that mean that the necessity of expensive, frustrating, inconvenient evacuations that turn out in retrospect to have been "false" is going to be a constant companion of coastal residents?

Yes, apparently so. Unfortunately, the situation is even more complicated than that.

Intensity: Will the wind pick up?

Intensity—the hurricane's power, as indicated by wind speed and barometric pressure and expressed as "category" on the Saffir-Simpson Scale—matters. (See Appendix.) What the government scientists would like most to improve at this stage is their ability to say if and when a hurricane will strengthen or lose power and thus change from one category to another. They want to know how strong the winds are not only in the upper reaches of the storm but also at the levels of rooftops. Stronger winds mean bigger storm surge, which means earlier evacuations of more people to save lives. Stronger winds also mean more destructive energy overall.

Damage potential increases exponentially as the square of the wind speed. Compare hurricanes of the same size: A Category 3 packs winds that do about twice as much damage as a Category 1 and pushes a wider and deeper storm surge. Incremental changes in wind velocities mean major changes in danger and in property losses.

To predict hurricane *intensity* change, the forecasters use a computer model different from those handling the *path* changes. Called SHIFOR (Statistical Hurricane Intensity Forecast model), it uses data about the hurricane's history, its current status and the climatological data that might influence it. Is anything in the air or in the sea that will heat up the hurricane or anything that will cool it off?

In 1998, two other computer models also were running the data to predict *intensity* changes, but these programs had yet to create any breakthroughs.[80]

"We can't forecast strength with any skill at all," Jerry Jarrell said in 1998. "We're not going to gain much now with track forecasting. We know as much as we are going to know about when and where.

"Our task now has to be to deal with the even bigger uncertainties related to intensity."

Since the data related to intensity is in the "top of the storm" rather than in the middle or the bottom of it, the scientists are grabbing hopefully onto data from the Gulfstream 4 jet.

Size: How big is that thing?

Along with *intensity, size* also is often one of the great unknowns and yet one of the most important factors determining a hurricane's overall impact. Everybody gets worked up about the "eye," the calm center of the hurricane, typically with a diameter of no more than 20 miles, and worked up about the "eye wall," the area of highest winds spinning around the "eye." But it is important to realize that hurricane power can extend outward hundreds of miles from the eye.

In addition to telling the public truthfully that they often don't know exactly what is going on inside a hurricane, the forecasters urge the broadcast media to quit showing maps with dots representing the hurricanes and lines representing the tracks.

Hurricanes are gyrating, spherical systems, of varying breadths and depths. Tracks are swaths, at least as wide as the distance between the extreme points of the gale-force winds.

Two fairly recent, familiar and powerful hurricanes illustrate the range of hurricane size:

- Category 4 Andrew's tightly wound core slashed a gash through south Florida, but the gash cut by the highest winds was only about 25 miles wide.

- Hurricane Gilbert, an enormous Category V, barely touched the United States, but it whipped up a territory in the Gulf of Mexico more than 500 miles wide after attacking the Yucatan.

The Great Sea Island Storm of 1893, a big one, produced 120-mile-an-hour winds in Charleston, South Carolina, more than 80 miles away from the eye south of Savannah, Georgia. At one point, it caused simultaneous flash flooding in the Midwest and on the Atlantic coast.

Size matters. A gigantic hurricane will wreck a lot more than a small one of the same intensity. From a large hurricane, the gale force winds might reach land 24 hours ahead of the eye. A large hurricane will push a storm surge over a wider chunk of the coast than a small hurricane. In general, the bigger a hurricane, the longer it will take the whole thing to move through a region, and the longer its victims will be under siege. If Gilbert had slammed into a major metropolitan area on either the Gulf or the Atlantic coastlines of the United States, how many lives and how much property would have been destroyed? Probably more than ever before in the nation's disaster history.

Forward speed

Add to the complex mix of hurricane *path, intensity and size*, the element of *forward speed*. Everything else being equal, a slow-traveling hurricane will pour down more rain on a given area, and spend more time wind-whipping that area, than a fast-traveling one. However, in many cases, a fast-traveling hurricane takes its hurricane-force winds farther inland than a slow-traveling one. In addition, a hurricane traveling landward at 24 mph would reach the coast in half the time in comparison to one chugging along at only 12 mph.

Moisture: How much rainfall?

What the forecasters know and what they predict about a hurricane's track, intensity, size and forward speed have the most influence on decision-makers and on the reaction of the public to the hurricane. In comparison to wind and storm surge, rain, which is much more familiar, might be considered benign. And yet, the amount of water a given storm will dump from the skies onto a given area tremendously affects the severity of the damage and suffering. In 1998, Frances was only a tropical storm, not a hurricane, when she flooded almost a fourth of the state of Texas. That same year, it was tropical storm Mitch's thirty-five inches of rainfall on deforested slopes that so devastated Central America and killed 9,000 people. Hurricane Floyd's rainfall, rather than the wind or storm surge, created North Carolina's $6.75 billion disaster in 1999.

From which direction?

One other aspect hard for the forecasters to predict but significant is the angle at which the hurricane's path intersects the coastline. A hurricane that runs parallel and close to the shore can inflict a lot of wind damage and beach erosion, and those living along the shore may get the impression that they took a hit. However, the most destructive hurricane is the one that throws its storm surge directly onto land when the eyewall itself charges inland at a more nearly perpendicular angle. Whatever happens to be in the hurricane's right forward quadrant will fall prey to that surging mound of ocean water.

Between the science and the coastal populations

The upshot is that while the scientists try to look at as many variables as possible during a hurricane threat, the public waits on shore for what the public hopes will be a simple verdict. *Is it coming here or not?* The forecasters document their historical margins of error: *Twelve hours in advance of the hurricane, the forecasters can say only that the hurricane will strike somewhere within a 100-mile stretch of coastline. Twenty-four hours ahead of a hurricane, the forecasters can say only that the hurricane will strike somewhere within a 200-mile stretch of coastline.* Considering the size of the planet and the vast amount of unknowns about its atmosphere, those accuracies are impressive. Considering the size of coastal populations, considering the vast amount of destruction and death hurricanes can render, consider the growing lengths of time it takes to evacuate coastal regions, however, that level of accuracy creates ghastly quandaries in evacuation timing.

It is scientists' job to give the best forecast they can for the storm. Period. Jerry Jarrell calls their forecasts "SWAGS" (scientific wild a…guesses). And yet, what they do and what they know, what they figure out and what they believe will happen have far-reaching impacts.

To help bridge the gap between the science in the National Hurricane Center and the state and local manager-political people who have to make the evacuation calls, emergency management officials working under the Federal Emergency Management Agency now serve as a *liaison team.* While the scientists fret over the many pieces of data that define for them the makeup and movement of the hurricane, the emergency people fret over what they know about the specific coastal towns and islands likely to feel its hot breath. While the forecasters work over the models and the information and analysis coming out of the models, the *liaison team* talks to the state and local officials who bear the weight of the decision on the possible evacuation of hundreds of thousands of people at a cost of tens of millions of dollars.

When the 1995 "hurricane from hell," Opal, was doing shenanigans over the Gulf of Mexico and the scientists were scratching their heads, it was the *liaison team* that sounded the alarm for action to protect the Florida Panhandle.

"They have become a key to our operation," Jerry Jarrell said about these non-scientists.

Hurricanes may be global, but protection from hurricane destruction and hurricane death is local, local, local, right down to the neighborhood and the homes in it.

Second-guessing

It's cool to make fun of the weatherman. *Yeah, the forecast was partly cloudy Saturday, and it rained us out of the championship game. So how can we believe them when they tell us what they think they know about a hurricane?*

Through the Internet, "unofficial" meteorologists—both professionals and amateurs—can access satellite pictures, radar data, sea-surface temperature data, ship and oceanic buoy data, tropical weather analyses and forecast fields, hurricane aircraft data, even the various official computer model forecasts. Links off the National Hurricane Center's own website promote more than a dozen kinds of hurricane-tracking software, although the Hurricane Center does not endorse any of them.

Those with the skill, the time and the motivation can work the same numbers the official forecasters work and come up with their own predictions.

Sometimes they agree, sometimes not. Occasionally, the unofficial forecasters get lucky and come closer to the truth than the official ones. But not often.

In most cases, unofficial forecasters doing their own calculating and predicting have vested financial interests in the evacuation decision. Like the official forecasters, they want to make the decision of least regret— least regret, that is, to themselves. Unlike the official forecasters, however, unofficial forecasters do not have responsibility for the public welfare. Mainly, if at all possible, they want their own business or industry to remain open and in operation. Hotels do not want their guests to go home, and paper mills do not want their employees to evacuate. Sometimes, the unofficial forecasters working for private enterprise are willing to take chances that officials responsible for many lives are not willing to take.

The more highly developed the community at risk, the more prone its leaders are to do "rogue" forecasting. In America everybody is entitled to public information and entitled to have something to say about it. Unfortunately, not all forecasts are of equal value.

Recalling an unnecessary evacuation started in Texas in 1988 because of an unofficial forecast aired over television, Robert C. Sheets, former head of the National Hurricane Center, warns: "Bad information going to the public can be dangerous. Interpreting the radar data, for example, is not as easy as it looks. Unless the TV meteorologists have taken extensive training to be able to do it, don't believe them."

Most coastal populations, businesses, media and officials rely on the National Hurricane Center. Of all the "hot seats" in the country, no seat is hotter than those in front of those computer monitors in Miami when a hurricane is headed toward this continent.

Even in the off-season, they face a myriad of challenges.

SLOSH delivers bad news

For more than a quarter of a century, a program named SLOSH (Sea, Lake and Overland Surge from Hurricanes), the National Weather Service's computer model for predicting likely hurricane flooding, has been scaring the mud out of the disaster specialists as well as lay people. SLOSH uses data about a section of the coast, its continental slopes, its shape, its inlets, its elevations, its populations, its highways. Based on assumptions involving dozens of theoretical hurricanes—varying in sizes, intensities, paths, speeds—the Weather Service runs the SLOSH models to tell where and how high the water will rise in a given community.

For areas inland, hurricane flooding from creeks and rivers is destructive and can be deadly. Storm surges do not create quiet swimming holes or friendly kayak trails. For the most exposed areas, including beachfronts, being slammed by storm surge is being slammed by hurricane-pushed ocean water. Storm surges—which vary in depth from 1 foot to more than 18 feet above the normal surface of the water—lay waste to buildings, split barrier islands and shift entire inlet channels. Not only do they come with fast currents and waves on top of them, in some places they come with massive quantities of debris—everything from wharf pilings to boat wreckage.

The coastal area maps produced under the SLOSH program show population densities and totals, show highways and their critical intersections, show the locations of public shelters for refugees. *Most importantly, they show water running and standing in places nobody wants water and in many cases places where hardly anybody believes the water will really appear.* Within a broad range, the SLOSH predictions about flooding from hurricanes have proven remarkably accurate, however.

SLOSH modeling is the basis of all evacuation studies in 36 study areas of the Atlantic and Gulf coasts and the Caribbean and Hawaii. The state of Texas, through Texas A&M University, has its own flood prediction program.

As critical to human safety as the SLOSH-based evacuation studies are, they are not done annually, nor even every five to ten years in all of the regions.[81]

Each study involves local, state and federal agencies plus engineering consultants, and each costs about $1 million. Budget constraints and the cumbersome process conspire against SLOSH studies timely enough to keep up with populations growing 75 percent in twenty years. Most coastal regions, with studies in the pipeline somewhere, face hurricane season after hurricane season with dusty, obsolete SLOSH data in their hurricane-response manuals.[82]

It's a wonder that more hurricane disasters don't degenerate into outright catastrophes.

On the other hand, not even perfect data, perfect computer modeling, perfect forecasting and constantly updated SLOSH maps could protect against hurricane fright, hurricane evacuations, hurricane damage, hurricane ruin and fatalities.

None of the scientific or technological advances in the 20th century keeps the Atlantic and Gulf coastal regions from being danger zones. As for the research and development constantly under way to improve the situation, Jerry Jarrell and his former Hurricane Center colleagues lamented that, if all the oceanography and all the technology were perfect, "Neither we nor anyone else can make the meteorology perfect."

Even more telling, forecaster Miles Lawrence admitted in a presentation for a recent state hurricane conference, that even with two satellites, Doppler radar, reconnaissance planes, the jet and computer modeling capability continually being upgraded, "We still can't keep up with the coastal populations."

In addition, even if both the science and the technology were perfect, and even if every coastal region were prepared to the maximum, no one could stop the hurricanes from coming nor prevent all of the damage they do. As important as the science and technology are today, no amount of brainpower directed onto a hurricane can change its brawn.

At this point, despite all the knowledge and the tools scientists have today, safety from hurricanes still depends on the skill of the coastal emergency management specialists and on the public itself. Those who do not get hit are lucky, and those who do get hit are unlucky, and nobody can change the luck of the draw. Nevertheless, those who get the information they need to live in safety in the danger zone and then act on what they know are far better off than those who would rather not be bothered by the bad news of hurricanes.

The scientists hope that the growing numbers of people on the nation's storm-swept coasts are coming to terms with their locations. They are in the paths of destructive, deadly monsters, still mysterious and uncontrollable. The forecasters know that over time, every area of the Maine-to-Texas coastline will be stricken.

If enlightenment of the coastal dwellers ever occurs, it will not likely happen because of data from the National Hurricane Center's reconnaissance jet, Gulfstream 4. More likely it will be a result of hearing the harrowing stories of coastal folk who have lived through hurricanes with water up to their armpits. Those who have witnessed the aftermath also can tell many a sorrowful story.

Chapter 6

Rescue, Relief and Recovery Today

'FEMA' roofs and expensive ice

"The universe is not hostile,
nor yet is it friendly.
It is simply indifferent."

John Haynes Holmes

Finally, in the modern danger-zone community that took the lick from the hurricane's hardest blast, the winds cease shrieking, the trees stop crashing and the electrical wires quit sizzling. The modern-day hurricane's survivors bring their dazed senses back to the scene. In only a few hours, everything has changed.

It is a different landscape, viewed differently through the eyes of every individual. Some see only the light in the eyes of people they love, miraculously still living after an unbelievably terrifying ordeal—and they weep for joy. Some see only ruins, and loss of lives, and they weep, too—for sorrow. The storm smashed one family's expensive new entertainment system. It crushed the bow of another family's shrimp trawler, a boat that had kept food on the table for ten years. The wind lifted the roof from a house, and water is standing in the living room. Soon the foulest odor imaginable will rise from the wet carpets. The terror from the sea tore up an apartment building, and the renters have no idea where they will sleep tonight, or next week.

In a once relaxed neighborhood, confusion and tension reign. Yesterday a home stood in this place. It was well known for homemade bread, piano music and roses at the door. Now it is a ragged mound of shattered glass, scrap lumber and wet Sheetrock. Where there used to be arguments over messy rooms, hugs over report cards, dreams about the future, there will be only questions today: *Why did this happen to us? What are we going to do? The place Dad used to work has been flattened. Mom, an emergency room nurse, has had only twelve hours of restless sleep in three days. It isn't fair.*

The stillness is oppressive. No mockingbirds sing. No hum comes from air conditioners or refrigerators. The tennis balls no longer fly back and forth over the net in the park—the rhythmical thwack, thwack, thwack, thwack on the court is gone.

In a short time, the chainsaws start growling, chopping their way through the trees lying every which way. Big tree trunks block streets. Tree limbs dangle from the traffic lights. Trees are plopped down in places they don't belong—on a broken porch, on the hood of a car, on the sliding board in the park.

Media on the scene

The helicopters soon begin to chop, chop, chop overhead. The media is there, taking pictures. The politicians are there, having their pictures taken as they survey the damage to their tax base and gauge the stress level of their constituents.

The battery-powered radio tells of a broken bridge and of the governor's request for presidential disaster declaration. It tells of a truckload of diapers and nonperishable food on its way to this place of disaster, on its way from some place where things are normal. *There are places in the country, even now, where the mattresses are dry, the TV is on and the stew is bubbling on the stove. It isn't fair.*

The radio urges listeners not to drink the water, not to panic, not to cut themselves with chain saws, not to get dehydrated in the heat, not to go into buildings weakened by wind and water—and not to set anything on fire with the lighting of the gas grills called into emergency service. Cooks are preparing to barbecue 14,000 pounds of meat, meat thawing inside thousands of freezers disabled by the thousands of trees that broke down the power lines.

For miles around, the electricity is off. Before long, the nausea will be overwhelming from the stench of rotting food, from the smell of decaying vegetation and from the odor of fellow humans who have been scared and who have sweated and who cannot bathe because there is no running water.

Did it have to be this bad? Didn't we have an emergency management officer? Did he forget to do something? What went wrong? How are we going to get help? How are we going to get going again? Who is going to put this back together? How? When?

Seems that for years these tropical storms hardly bothered us here. Some of them didn't even turn into hurricanes. Some of them moved on out to sea and we never heard of them again. They always used to go somewhere else. On TV, we saw newscasters standing with their microphones in the rain and wind blowing their umbrellas, with the surf rolling behind them. We paid attention for a little while, even evacuated twice in ten years, but we got used to the idea that hurricanes don't come where we live.

And now this.

There is nothing "fair" about any of it.

And yet, here it is. This one's for us.

The evacuation and battening down have drained the sap out of the bones. Sleep has been scarce and fitful. Hyperactivity turns suddenly to exhaustion. A toilet has landed in the street. *Isn't that a hoot?* Everybody laughs. *But the toilets in the bathrooms left standing do not work.* There is nothing funny about non-flushing commodes. The work of responding *intuitively* to the emergency and then *sensibly* to the recovery has only just begun.

In the middle of the chaos, appeals for response to the hapless hurricane victims begin to produce results. A spontaneous donor/helper brings plastic sheeting to keep the rain off the breakfast table. Volunteers show up with chain saws and Gatorade. Ice and bottled water begin to arrive. A young woman brings boxes of cookies, apples for the children, canned tuna. The menu for supper in the public shelters will change shortly from cold bologna sandwiches to warm chicken soup.

To the rescue

The Federal Emergency Management Agency has spent a lot of man-hours trying to prepare for this one. Now it is going to be tested.

FEMA has many jobs, but after the hurricane, FEMA's main job is to look hard at the big picture and to summon the resources of the federal government to help thousands of storm victims who cannot help themselves.

Clearly, somebody needs to assess the damage, and somebody needs to get the fallen trees, the battered sailboats and the building debris out of the roads. Somebody needs to restore the power and the telephones. Somebody also needs to sound the warnings about the frayed electrical wires, the poisonous snakes that crawled into houses during the flood. The sewer and water systems have to be flushed and fixed and certified.

FEMA is the link to the National Flood Insurance Program, through which flood-prone properties have to meet certain requirements and through which they qualify for certain significant subsidies for insurance in the flood-prone zones. In the aftermath of hurricane destruction, FEMA is simultaneously praised for responding to the emergency and lambasted for having encouraged, through its flood insurance program, development of the storm-swept places.

Through FEMA, after the storm nevertheless, federal officials also can give limited amounts of money and lend limited amounts of money at low interest rates to individuals, businesses and state and local governments. FEMA sets up shop to offer assistance and information in person and sets up toll-free numbers. To prevent post-disaster disaster, the mayor and the governor and the congressman need to give out the same information, need to sing in harmony. Despite good will and good intentions, discord and mixed signals tend to take over. FEMA representatives try to give all public officials the same sheet of music.

To make what would otherwise be unlivable housing livable temporarily, FEMA provides plastic tarpaulins, usually royal blue in color, often called "FEMA roofs."

Red Cross

As the agency chartered by Congress more than 100 years ago to be the nation's chief volunteer disaster relief agency, the American Red Cross has spent a lot of man-hours in the 20th century dealing with hurricanes. Professional staff and trained volunteers rush to the wind- and water-whipped places, bringing with them the ability to summon assistance from many sources. Volunteers specializing in nursing, mental health and social services come, first from the affected state and then, if necessary, from all over the country. It is tough work. They go about it with welcome expertise.

Pat Goodale, Red Cross disaster specialist, served as night job director in the Hurricane Andrew relief-and-recovery program in Florida, among other tours of duty. "It is unbelievable what we have to ask (the volunteer) people to do—and that they so willingly do it," she said.

"Sorry, but we have only one motel room for four volunteers. It has two beds in it. Can you sleep in shifts?

"No toilet facilities are working in this facility. Here is a baggie.

"Yes, the carpet is wet in that interview room, but we have nowhere else to work just now.

"If you have to take down information on a case in the dark, and there is no power, we suggest you pick up a flashlight. They are stored in that box."[83]

If the destruction is widespread and the impacted population large, the emergency shelters will become impact shelters. Instead of feeding the masses for 48 hours, the Red Cross will take on the task of feeding as many as come to the door hungry for as long as they keep coming. Clara Barton started the compassionate policy in 1893 after the Great Sea Island Storm, and it continues today. Nobody asks about money in the food line. There is no means testing to get a place in a Red Cross shelter to lay down a sleeping bag.

Usually, within 30 days most disaster victims have been able to find places to eat and sleep other than public shelters. With so much in disarray, getting back home, or at least into *a* home is a high priority. However, after Hugo, some folk whose dwellings had been wiped away in the storm were hopeless as well as homeless. They had never been in such predicaments before. For them, there was nowhere in the region to live. With the inexpensive housing stock smashed, they could find neither apartments nor mobile homes nor small cottages to rent or buy at prices they could afford. While they scrambled, trying to make arrangements, the Red Cross provided food and living quarters for some of them for three months.

The Red Cross spends millions of dollars on hurricane relief and recovery, money that comes from private contributions. Hugo cost the Red Cross $63 million. Andrew cost $83 million. When a crisis occurs, the opportunity to donate is the first message on *www.redcross.org*.

Cooking in a truck

After Hurricane Carla ripped through Texas in 1961, Southern Baptists rushing in to help found Carla's survivors nervous, confused, weary, hungry—and unable to buy, keep and prepare food. So the Baptists set up mobile kitchens. Hurricane Beulah in 1967 brought even more Baptist volunteers out, and the next thing the Southern Baptist Association knew, it, like the Red Cross, was deep into the business of disaster relief. In 1998, the Baptists had 200 "disaster relief units," a term which means mobility, equipment, supplies, and all together more than 15,000 trained volunteers.

A pattern developed in which the U.S. Department of Agriculture and the Red Cross would provide the food, and the Baptists would cook it. Volunteers went to mass-feeding school to learn to prepare up to 10,000 meals a day.

They operate out of truck kitchens, which they drive to disaster scenes all over the continental United States, and out of airlift kitchens, which they fly to disasters in the Caribbean and Hawaii. In the last thirty years, the Baptists believe they have prepared about two million meals for people whose lives have abruptly changed for the worst.

At the same time hurricane survivors need food, they need other basics. So the Baptists formed volunteer "chain saw units" to clear roads and driveways of downed trees. They created volunteer-staffed "communication units," basically travel trailers equipped with radios, computers, telephones. They have organized "child care units." They also provide "shower trailers," which, when attached to pumper trucks, offer welcome showers to people grimy and weary from dealing with the aftermath of flooding and wind damage.

One of the kindest, nastiest tasks the Baptists do for people whose homes have been flooded is to go through them painstakingly, mucking out the mud and mire, sanitizing the floors and walls and whatever else can be salvaged.[84]

Along with the Red Cross and the Baptists, hundreds of churches, the Salvation Army, other organizations and ad hoc groups spring into action in the wake of hurricane strikes in the United States. All of them depend heavily on volunteers and on willing donors. The generosity of time and of dollars often begins to take shape quickly, even spontaneously.

And yet, addressing the immediate needs of hundreds of thousands of stressed hurricane survivors—as complex and physically exhausting it is—is far easier than addressing the long-term problems. Quick relief is a stroll on the beach compared to recovery.

For most people, most of the time, how they get their lives back together after a hurricane hits them depends largely on *themselves.*

Those with few resources before the storm find it hard to get going again after the storm destroys what they had. The truly destitute can get emergency grants for necessities, but then come multiple problems requiring longer-range solutions: *Where will I live now, and how can I sustain myself? Where will I work? How close am I to my credit card's limit?* For some, low-interest Small Business Administration loans can help; but the loans have to be paid back, often under difficult circumstances. Add to the overall problem the lesser problems of assembling again the trappings of everyday life that have been blown away—everything from bedding to calendars to soup ladles and kitchen knives to hairbrushes—and hurricane survivors often feel so overwhelmed they cannot function.

Those well insured and well off financially before the hurricane can become almost whole again in time. In many cases, insurance money that pours into a community after a destructive storm makes possible renewal that would have been impossible. If the house needed a new roof anyway, and the storm caused the insurance company to pay for it, the storm victim might come out a winner, at least in that regard.

Between the wealthiest and the poorest, moderate-income coastal dwellers often carry the biggest personal burdens of all. It can be a shock to find oneself underinsured, unemployed and nearly homeless at the same time. It also can be a shock to realize how interdependent we all are.

Even the most well-prepared, self-sufficient families suffer indefinitely if others in the community—the local government, the businesses, the nonprofit agencies and other families—cannot pick themselves up, dust themselves off and start all over again soon. Individual homes are linked to one another by shared resources such as electrical wires and streets, public water and sewer service, law enforcement, medical service, schools, churches and the availability of necessary goods. In ways few understand until they have been hurricane-whipped, a lot of every individual's well being links directly to the whole community's well being.

In recent years, three states' hurricane experiences illustrate the problem.

The Hugo aftermath in 1989

"When Hugo hit South Carolina," two academic researchers wrote afterward, "the warning and evacuation activities worked well, but, when the storm was over, *no one seemed to know what to do next*. Unfortunately, very little post-hurricane planning had been done at any level of government. The recovery had to be improvised."[85]

As Hurricane Hugo had significant impact on more than 100,000 families in 24 counties, the sheer scope of the problem made it hard for even skilled, well-meaning people to deal with it. No experience was available to lead the way because the last hurricane to strike South Carolina hard had been Hazel in 1954—thirty-five years earlier.

After Hugo, everything that could go wrong went wrong, at least once.[86]

Tempers flared. Both FEMA and the Red Cross, already stretched because of Hugo damage in the U.S. Virgin Islands and Puerto Rico, took a while to gear up to handle Hugo's catastrophic damage in the United States. FEMA did not get its field office open in South Carolina until eight to ten days after the hurricane. U.S. Sen. Ernest F. Hollings, D-SC, publicly assessed FEMA as "a bunch of bureaucratic jackasses."[87]

Confusion mounted. At one point, South Carolina after Hugo had two information networks, one through the emergency preparedness division and one through the office of the governor. Communications broke down at every level.[88]

Officials overreacted. Charleston Mayor Joe Riley appealed over and over through the media for people to send food and clothing. Although 24 (mostly rural) counties were in dire circumstances, truckload after truckload of supplies congested highways leading directly to Charleston itself. Then, relief workers had to figure out how to move massive amounts of groceries from where they were not needed to where they were needed.[89]

Haste caused problems. Various nonprofits, church representatives and groups that rallied suddenly rushed in to provide repairs and rebuild damaged homes. Some structures were put back in a better condition than they were in before. Some were put back without regard to even the minimal existing regulations, including FEMA's National Flood Insurance Program.[90]

The Andrew aftermath in 1992

In the more massive Andrew disaster in Florida three years after the Hugo disaster, even more victims suffered keenly and for a longer time. There, as in South Carolina, the warning and evacuation to prevent fatalities worked. After the winds died down, however, communications, relief and recovery fell into a black hole.

The state of Florida was not prepared to handle the impact of the hurricane on its people. "There was little coordination between the State Division of Emergency Management, county emergency management offices and the Federal Emergency Management Agency.... The absence of pre-established coordination procedures between all three parties, as well as a philosophical reluctance of all of the organizations to work in concert, resulted in a confused and much less timely response in addressing the immediate and long term needs of the impacted areas.

"The state had to rely on broadcast news media for information regarding damage to the Homestead area," a Florida official said candidly.[91]

Fortunately, FEMA, having caught a lot of flak and learned a lot of lessons from Hugo, took new ways of doing a few things to Florida. The agency tapped every disaster-experienced staff member and volunteer available. And yet, criticism of the way FEMA handled Andrew's aftermath was so sharp and so persistent it spurred FEMA afterward to a much more aggressive role in hurricane planning.

The Red Cross also had gone to school on Hugo, and its volunteers hurried to Homestead freshly experienced in dealing with the kind of mess they would find there. Only there the mess was bigger because the hurricane had barreled into a more densely populated region. Situations over which the trained staff and volunteers had no control—shortages of water, food and staff, and damage to shelters themselves—hampered and frustrated them. The work went on day and night nevertheless until every storm victim had a better alternative than dependence on temporary solutions.[92]

As an example of the spontaneity that made a difference after Andrew, Charleston Mayor Joe Riley promptly dispatched experienced help from South Carolina to Florida. Within hours after the hurricane hit, the Charleston traffic and transportation director began assisting in the clearing and repair of roads in Dade County. Within only 12 hours after reaching the area, thirty-five Charleston policemen made 95 arrests for looting. After breaking into a junior high school to create a temporary emergency medical center, in their first 24 hours of operation in Florida, Charleston-based paramedics treated 5,000 storm survivors for cuts and bruises, nausea and diarrhea from drinking unsanitary water.[93]

Hundreds of semi-trailers, pickups and vans poured into south Florida, carrying canned ham, bottled water, baby formula, paper plates and the other commodities. Supermarket chains donated cash, water, rice, bread, rolls and muffins. Drug store chains donated anti-infection drugs, insulin, eyeglasses. Fast-food organizations set up mobile restaurants and served free food.

For several weeks, in typical but heartwarming fashion, generous donations from thousands poured into the region. For a few weeks, residents whose own personal lives had been shattered reached out to one another, helping, giving, encouraging, and sharing. For a few weeks, businesses facing tremendous financial burdens themselves fed the hungry, gave goods and services to the needy and created new temporary communication and transportation systems.

Despite the scope of the catastrophe, and the horror of the suffering it caused, as far as official sources know, nobody died of starvation or malaria or exposure because of Andrew.

Still, Andrew brought on suffering and insolvency to a lot of people unprepared to fend for themselves. Even more so than Hugo, Andrew forced what had become a complacent public to stare at the truth of the serious, long-lasting misery that one hurricane can create.

For four decades, new settlers had poured into Florida, and since 1980, the coastal population of the state had increased almost 37 percent. These settlers' new homes looked OK when they moved in, but the good-looking exteriors covered structures too frail to make it through the hurricane. The places of business sprung up to support the new residents also were built inappropriately for a region known to have experienced such high winds and massive flooding in the past. Officials had worked together on the critical problem of evacuation of people but had not planned adequately for their community's infrastructure, public and private, and certainly had not planned adequately for the next problem—what to do in the aftermath.

More than a half dozen years after Andrew hit Homestead, the community still bore not only the scars but also some of the wounds. The post-storm situation was wretched, and it did not have to be as bad as it was.

Bertha, Fran and Floyd

Four years after Andrew, in the summer of 1996, North Carolina got two "big blows." Hurricane Bertha, a nasty Category 2, swatted first. The Tarheels had barely gotten over Bertha when eight weeks later Hurricane

Fran, a Category 3 at landfall, punched the same beaches again—but harder, with an 11-foot storm surge and sustained winds of 115 mph. More than 30,000 homes, almost 900 businesses, timber and agricultural products suffered $6.57 billion worth of damage. President Clinton declared a disaster area taking in 51 counties.

Casualties from Fran included the collapse of 90 percent of the beach houses on Topsail Island, a strip of sand identified by Congress in 1982 as a Coastal Barrier Resource Act property, unsuitable for development and therefore unsuitable for federal assistance, including federal disaster assistance. One state legislator looking at the Topsail Island wreckage from the air commented: "We need to get out with the least amount of cost possible and (not) come back. You just can't live in certain parts of the coast. They're not made for homes."[94]

Returning to Holden Beach after Hurricane Fran, the town manager said, "I know there were houses in those spots before I crossed the bridge. There were just septic tanks there today."[95]

By the summer of 1996, the cadre of hurricane-savvy volunteers had expanded. After Fran, hundreds of victims of Hugo from South Carolina and of Andrew from Florida poured into North Carolina to help. They chopped wood, repaired roofs, hauled trash, served food and commiserated. Most people in the region waited days for the power to be restored. Some waited several weeks. "If not for the heroic efforts of out-of-town volunteers, the wait would have been much longer," according to the Wilmington Star's after-the-storm publication, *The Savage Season*.

Using the experience in its worst hurricane season in memory, North Carolina's Division of Emergency Management radically reorganized itself early in 1997. The first message out was to individual families: Prepare for at least three days of self-sufficiency after a hurricane. After that came initiatives to figure out what resources state and local governments would need before, during and after a hurricane; to plan for their deployment and to keep track of them by computer.

North Carolina's hurricane program planner expressed pride in what the state did. Then he remarked: "Can you ever be 100 percent prepared, or can you only be constantly preparing?"[96]

In 1999, North Carolinians' constant preparation was tested yet again as Hurricane Floyd swamped the eastern one-quarter of the state. "The ground is too soggy to bury the dead. Entire farms of tobacco and cotton have disappeared. Dead hogs and chickens bob along with kitchen chairs and coffins in the filthy floodwaters. 'There's no doubt there are individuals out there who were swept off roadways and whose bodies have not been recovered,' said the state's chief medical examiner."[97]

For danger zone residents everywhere, constant preparation—by individuals, businesses and public officials—offers the only hope for safety. As the Fort Lauderdale *Sun-Sentinel* says, get your plans laid and your work done during the "lull between storms." Coastal communities are growing and evolving. Community leaders continually change. After a hurricane makes shambles of a place, it is impossible to make up for a lack of foresight to create a post-hurricane program ahead of time.

A lot to sort

After the 1893 hurricane in South Carolina, one of the Red Cross volunteers working for Clara Barton said, "Everything needs to be done at once." At that time in that place, it was essential that people be given grits, be given shovels to dig ditches to drain the standing water and be helped to put new food crops in the ground.

In this more complicated era, along with mounds of ghastly physical debris, everything after a hurricane hit gets all tangled up—the authorities, the jurisdictions and the priorities. Which ruined asset should be tackled immediately, and who is going to decide? Where lies the authority? Who will handle it? Who will be told? How? When? Who will pay?

Where, along the 104 miles of prostrate power lines, should the linemen begin work immediately?

The children need to get back to school, but how can a teacher teach math when she has no bed to sleep on? And where can the class be held when the windows of the classrooms are shattered? Insurance companies need to disburse checks, but how can they assess the damage when the landmarks and street signs have been obliterated and when there is nowhere for the insurance adjusters to find lodging?

Food must be delivered, but where must it go, and who will distribute it? Where will the ice come from, and the drinking water, and the insulin, and the baby formula? People need to get back to work, but how can they go to their jobs when the offices and shops and hotels they worked in are caked in mud or demolished? How, oh how, will all of those tons of debris be removed? And where, oh where, can it be taken? What permits are required? How can all that must be communicated *be* communicated? How can all that must be transported *be* transported?

At times like these, public health and safety issues must rise to the top. There must be some rationale for clearing one street ahead of another. When storm victims believe that wise heads are "in charge" and that in due time the multiple pressing needs will be met fairly, most can make progress on their own. However, when individual storm victims sense that no one is really "in charge," or that decisions are being made willy nilly, or not being made at all, they have a hard time focusing on their own part to put their lives and put their community back together. Worse, they can fritter a lot of time away worrying and complaining.

Despite documented evidence of the importance of thorough post-hurricane plans for every coastal community, few have them.

The critical issues of relief and recovery got a lot of attention in South Carolina after Hugo, even more attention in Florida after Andrew, and concentrated attention in North Carolina after Bertha and Fran in 1996 and after Dennis and Floyd in 1999. Experience certainly might pay off in future disasters in those states.

On the other hand, the odds are against the experienced hands in South Carolina and Florida and North Carolina being in charge for the next hurricane disaster where they live. Three or four tropical storms or hurricanes make landfall most years, but they choose the particular shorelines randomly.

Sad to say, human nature makes it hard to use the experience of others to learn—especially to learn some of the most valuable lessons of all.

About the illustrations

In ways that words can never achieve, images convey the power, the pathos and the paths of hurricanes throughout history.

The following small album represents thousands of pictures, maps and graphics that depict the random nature of these storms—and the kinds of tragedies they have delivered various communities over time.

Source: Meyer Schein
Downtown Beaufort, South Carolina after the Great Sea Island Storm of 1893

Figure 1

Source: Rosenberg Library
Burying Bodies in Galveston, Texas—1900

Figure 2

Source: Leslie L. Tyler Engravings
Connecticut Shore after the New England Hurricane of 1938

Figure 3

Source: NOAA
Monster Hurricane Gilbert—September 1988

Figure 4

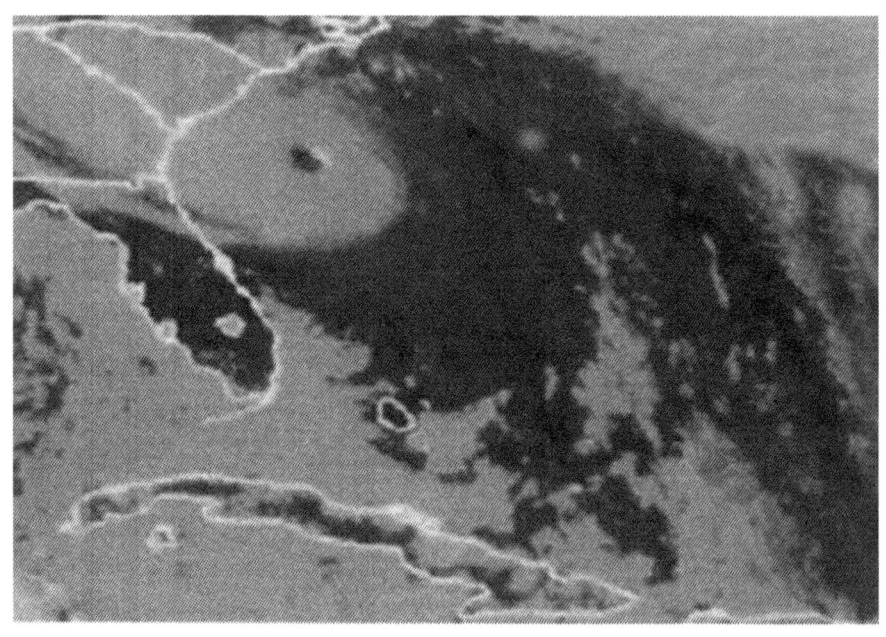

Source: NOAA
Hurricane Hugo Heads for South Carolina—September 1989

Figure 5

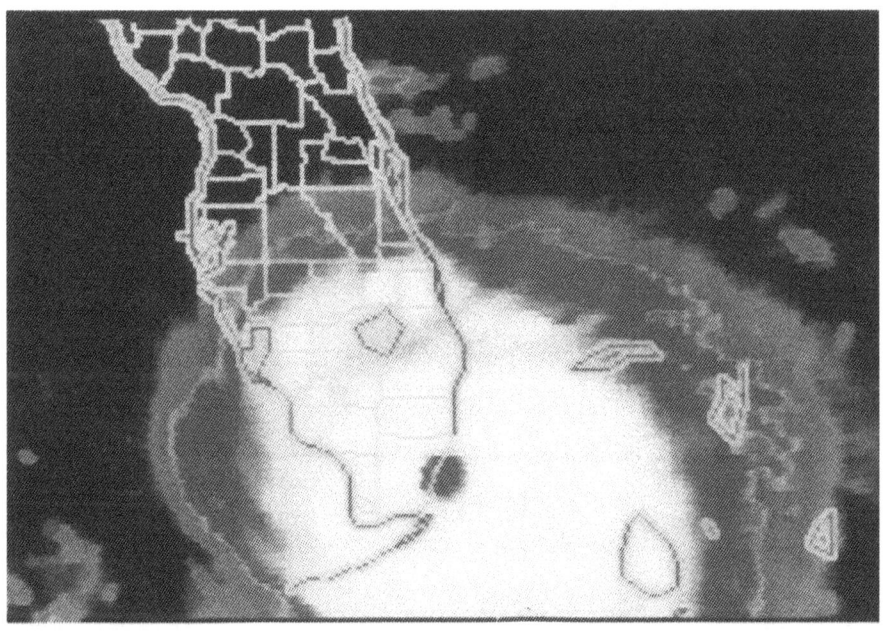

Source: NOAA
Hurricane Andrew at Miami, Florida—August 1992

Figure 6

Source: NOAA
Hurricane Andrew Damage in South Florida—August 1992

Figure 7

Source: FEMA News Photo
Destruction by Hurricane Opal at Panama City, Florida—1995

Figure 8

Source: FEMA News Photo
Hurricane Fran Damage on North Topsail Island, North Carolina—1996

Figure 9

Source: NOAA

Caribbean Birth of Hurricane Mitch in October 1998

Figure 10

Source: National Guard Bureau
Army to the Rescue after Hurricane Floyd in Pollocksville,
North Carolina—1999

Figure 11

Source:FEMA News Photo
Mobile Home Damage from Hurricane Bonnie in Virginia—
September1998

Figure 12

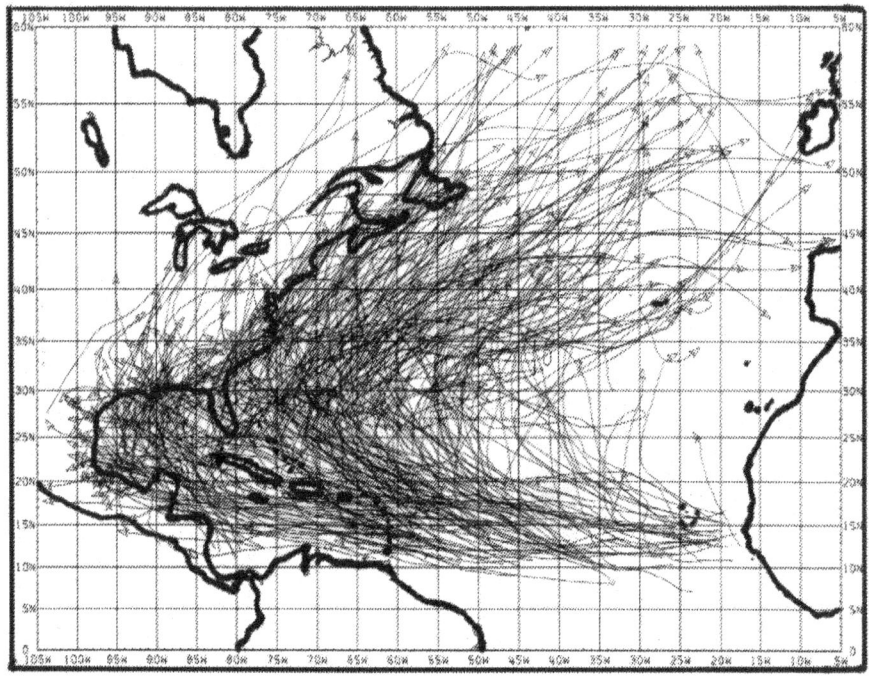

Source: NOAA
Tropical Storms and Hurricanes Beginning in September 1886–1992

Monthly Totals of Storms or Hurricanes 1886–1992

June	56
July	68
August	217
September	308
October	189
November	42

Figure 13

Chapter 7

'Did it Have to be this Bad?' 'No'

Hold the roof on the house

"A good scare is worth more to a man than good advice."

Edgar Watson Howe

Early in the 21st century, about 50 million people live in communities that will be wrecked when hurricanes slam into them.

From those who personally escaped a hurricane's terror but returned to find their homes turned into rubble comes a sorrowful if shrewd observation: "The view was magnificent, but what a price to pay. Geez! What went first, the roof or the doors? They didn't tell me about all this when I moved here." Actually, the hurricane history in the danger zone is readily available for those who ask, but few ask. In 1998 alone, Americans built 50,000 new housing units on the barrier islands, the most hurricane-vulnerable properties of all.[98]

The Coastal Barrier Resource Act of 1982 designated 186 strips of low-lying coastal lands as ineligible for federal flood insurance. The law has not lived up to its promise, unfortunately—in part because the very wealthy have found those properties attractive simply because they are so unavailable to the masses and have built high-priced houses on them, creating exclusive, private communities. Also, of course, the law does not apply to great swaths of properties under development at the time it was passed. So it has not effectively discouraged growth in some of the riskiest spots along the shore.

From mobile home owners hit by a hurricane, most of them inland rather than on the more expensive waterfronts, often comes an even sadder lament: "We've lost everything." Mobile homes have proliferated near the coast. They accommodate retirees drawn by the appeal of beaches and mild climate but unable to afford conventional housing. They accommodate workers drawn to the coast by jobs but also unable to afford other dwellings. Although some "manufactured housing" is far better built today than in the past, those living in mobile homes, double-wides and "modular units" in coastal counties often are told to evacuate for every hurricane headed their way. The photographs of the wreckage from past storms are enough to get most of them moving.

It is a shame that so much of the coastal housing stock cannot resist hurricanes.

Out of the recent workshop of experts called together after the Hurricane Floyd experience came this powerful admonition: "The Forum *strongly* recommended that governments at all levels (federal, state and local) *urgently review and strengthen building code and land use policies and enforcement* with the explicit intent to ensure the maximum possible survivability of existing structures and infrastructure."[99]

In general, residents of the danger zone pay more attention to the wallpaper in master bedrooms than to the ability of the house to hold itself together in a hurricane.

In general, building practices in coastal regions do not create hurricane-resistant homes, businesses, schools, courthouses and libraries. The codes have improved, but they do not necessarily do the job. Even where the codes are strong enough, in general the building inspectors do not inspect precisely and thoroughly enough to make sure the codes control construction practices. In general, when building materials claim they can withstand certain wind speeds, they have not been tested against hurricane conditions.

And yet, with changes in the culture of every coastal town, it would be possible for coastal dwellers to resist a lot of the destruction hurricanes bring when they come.

In the most vulnerable places, it is the dramatic storm surge that drowns people and destroys structures in its way. When more than 10 to 12 inches of rain falls in a few days, flooding also endangers lives, wrecks property and renders communities unlivable. Close to the beach, near a coastal creek and adjacent a low wetland, homeowners now know they can expect ocean flooding in hurricanes. Across the country, 19,000 communities, both in river valleys and along the coastlines, have taken the steps of mapping and passing laws to become a part of the National Flood Insurance Program. Simply building houses on sturdy, well-anchored pilings above the height the water is most likely to rise has paid off for tens of thousands of people. When such houses also are designed and built to handle high winds, they sometimes make it intact through raucous weather.

However, even elevation cannot save a fragile mobile home inadequately anchored or a house inappropriately located, unwisely designed or poorly built. On many beach houses, for example, vast expanses of glass invite hurricane wind and rain destruction. The elaborate, many-gabled roofs issue the same invitation.

When families find their home uninhabitable after a hurricane, the most common culprit is the wind. As measured in dollars, the damage in most hurricanes is wind-driven, according to the insurance companies.

As hurricanes, or even gales, are raging, they wreck buildings in many ways, but this process is typical: A whipping wind works its way under a single roof shingle. From that vantage, it pries up more shingles. Suddenly, like a blocker moving down a football field to make space for a ball carrier, the wind opens a new route into the house. The hurricane drives the rain into it.

Imagine a high-pressure fire hose pouring water into the entire house for several hours, and you get the picture of what a relentless wind can do if it can pry open one hole, whether through the roof or a door or a window, and then use that hole to pry open an even bigger entry.

Away from the waterfront properties, wind-resistant construction is rare. The untested theory much bandied about has been that hurricanes lose half of their intensity about 12 hours after leaving the ocean. That does not always happen, however. Sometimes winds pound unmercifully on communities well inland, communities not likely to be evacuated but very likely to be beaten into rubble.

Before the hurricane-force winds of the Great Sea Island Storm of 1893 subsided, they created a path of destruction from coastal Georgia and South Carolina all the way to Pennsylvania, New York and New England. Hugo struck with a terrible force and then charged rapidly inland in 1989. One hundred miles from its point of landfall on the coast, Hugo wrecked Sumter County, South Carolina, to the tune of $800 million. One hundred fifty miles inland, in and near Charlotte, North Carolina, Hugo's winds wreaked more havoc, creating ten years' worth of "green waste" (from trees) in three hours.[100]

So how far inland from the ocean does one have to live to discount the potential of hurricane winds hammering at the rooftop? That depends on the amount of risk one is willing to take.

For any construction within 50 miles of the Atlantic or Gulf shoreline, it makes sense to apply tougher than ordinary wind-resistance standards. They are not onerous. They are appropriate not only for personal decision-making but also as a communitywide commitment, enforceable by inspection. One might benefit from building a really sturdy house for one's own family and—and then find that a shoddily built neighbor's house breaks up in a storm and becomes a missile source. Even a single exploding structure can send lumber and cracked masonry flying all over the neighborhood. Hurricanes have been known to drive two by fours through tree trunks.

Who knows the most about protecting property in the danger zone? Insurance companies, of course, although they have come only lately to their knowledge.

Insurance industry quaking

In the mid 1980s, as coastal populations swelled, insurance companies began to worry about the amount of property they had insured in the hurricane-vulnerable areas of the country. Through the Insurance Research Council they analyzed the financial impact of a "worst case scenario," two imaginary disasters back to back, each one doing $7 billion worth of imaginary damage to insured properties.

In 1986, the industry's study concluded that, if two such disasters should strike this country, the insurance industry would go broke.[101]

Then came Hurricane Hugo to South Carolina in 1989, demolishing the waterfront properties and ravaging the inland properties in its wide path through the Carolinas, Virginia and Pennsylvania. It was a "$7 billion hurricane" whose insured losses totaled $4.2 billion. The bulk of the insurance claims came from people well inland from the ocean—for damage *started by the winds and exacerbated by rain.* In a single day, the South Carolina Wind and Hail Underwriting Association lost more than its total revenue in its eighteen years of existence.[102]

Three years after Hugo, Andrew hit Florida and Louisiana. The "$30 billion hurricane" whose insured losses totaled $15.5 billion, Andrew also showed what *wind* does to casually constructed communities. In only a few hours early on the morning of August 14, 1992, in south Florida, more than 28,000 homes were destroyed; and another 107,380 damaged, their roofs lifted, their interiors flooded, their unreinforced concrete block walls rammed with trusses ripped from their neighbors' houses.

Flying debris had slammed into more than 160,000 vehicles. More than 82,000 businesses had been wind-whipped, water soaked or destroyed altogether. At least 11,800 mobile homes had been wrecked, a high percentage of their owners filing claims for total loss of structures and contents.

Abruptly, Andrew, mostly through its wind, made 180,000 Floridians temporarily homeless and 75,000 Floridians temporarily jobless.[103] In many ways, for a short period of time, their predicaments were as horrible as the predicaments of the "sufferers" ninety-nine years earlier—after the Sea Island Storm blew away the houses, the tools, the crops and the phosphate dredges in South Carolina.

One difference between the aftermaths of hurricanes past and Hurricane Andrew is that to 370 property-casualty insurance companies, Andrew meant the investigation and payment of more than 680,000 claims.[104]

Like thousands of families laid financially flat by previous hurricanes, nine property-casualty insurance companies lost their solvency as a result of Hurricane Andrew. Instantly, with insurance pools having to pick up the pieces, the insurance business as a whole found itself in turmoil.[105] Like a single hurricane victim asking himself amid the rubble of what used to be his home, the insurance companies wanted to know, "What went wrong?"

How much at risk?

After reeling from the claims from Hugo, Andrew and Hurricane Iniki (Hawaii, $1.6 billion), plus the losses from the Northridge earthquake in California, the property insurance companies directed the Insurance Research Council to tally up their collective exposure on hurricane-vulnerable properties.

Before the study, the dollar figure loosely bandied about was $2 trillion. But the $2 trillion turned out to be low by more than one third. Insured exposures for just the counties fronting the Atlantic and Gulf coasts totaled *$3.15 trillion* in 1993.[106]

Even more alarming than the total itself were the rates of increase. Between 1988 and 1993, insured property values in the coastal counties of all the Atlantic and Gulf states went up more than 11 percent a year.[107] If the values continued to climb at that same pace for the next seven years, by 2000 the insured exposure in the most hurricane-vulnerable counties would be about $6.6 trillion.

Two facts compound the problem of the potentially exorbitant total price tags on future hurricanes: 1) The pace of coastal growth for several decades has been faster in the Southeast and along the Gulf coast than in the Northeast. 2) The Southeast and the Gulf coast are the most prone to hurricane strikes.

The insurance companies' fear was not so much that bigger and wilder hurricanes might strike these areas harder—although some experts believe that could happen due to global warming and sea level rise. Their fear was that what happened with Hugo and Andrew and Iniki will happen again, and worse—that the same kinds of hurricanes will damage more and more insured property simply because more and more is being built.

On the basis of hurricane history, the insurers estimated what their "expected losses" will be for the major coastal regions. They used 1993 figures and have not updated their projections since. [108] In the only such comprehensive study the insurance companies have done, they estimated conservatively that at least every 50 years hurricanes will cost them $16.9 billion in the Southeast, $7.3 billion in the Northeast, $6 billion in the Gulf region and $2.5 billion in the Mid-Atlantic states.[109]

The insurance companies also used the 1993 values to look closely at key points along the coastline and estimated insured damage totals if they should catch the brunt of Category 4 and Category 5 hurricanes. The damage potential of a Category 5 for their test cases is as follows: Miami, $52.5 billion; Ft. Lauderdale, $51.9 billion; Galveston, $42.5 billion; Hampton, Virginia, $33.5 billion; New Orleans, $25.6 billion. The damage potential of a Category 4 for their test cases is as follows: Asbury Park,

New Jersey, $52.3 billion; New York City, $45 billion; Long Island, $40.8 billion; Ocean City, Maryland, $20.1 billion.[110]

The Insurance Research Council's cost projections, as high as they were, took into account *insured properties* only. Computing the total expense of future hurricane damage requires adding on at least the following:

- The costs of other uninsured losses.

- The costs to the publicly subsidized flood insurance program.

- The costs of repairing and replacing with tax dollars significant amounts of public infrastructure—such as roads, schools, land-fills, etc.

- The costs of restoring the functions of utilities—electric, water, sewer, cable TV, etc. (costs passed on to families and businesses).

- The costs associated with the endless lists of out-of-pocket expenses for individuals, businesses, charities and local towns and counties, of everything from temporary housing to overtime pay.

The whole truth is that the total expense of hurricane damage probably has never been calculated, simply because too many individuals, businesses and institutions have borne the costs—and even they would have a hard time tallying up all of the money they had to spend.

Out of the Forum on Policy Issues in Hurricane Preparedness and Response, developed by the American Meteorological Society and sponsored by The Weather Channel in June 2000, came this prediction: "Andrew and Floyd do not approach the catastrophic losses of life and property the nation will face when a major hurricane makes landfall at a large population center such as Miami, New Orleans, Tampa-St. Petersburg or New York City. Death tolls could once again be in the thousands, figures not seen since the Galveston hurricane a century ago or the Lake Okeechobee, Florida hurricane of 1928. Economic losses could exceed $100 billion. It could take up to six months to establish basic infrastructure. Overall economic recovery could require decades."[111]

What's going on?

Jolted by Hurricane Andrew into the reality of what hurricanes can do to them, the insurers sent adjusters and others peering into battered walls and ragged doorframes while they still lay on the ground just north of The Everglades.

"Did it really have to be this bad?" insurers, the media, civil engineers, homeowners, the construction industry and government asked one another as they look around at the unbelievable wreckage. Most observers believed the answer was "no."

Before Andrew struck, South Florida had one of the strictest wind-resistance building codes in the country. So what went wrong?

The Insurance Research Council concluded that a 40-year "false sense of security" about hurricanes had led to "helter-skelter" development, "lackluster code enforcement, (inadvisable) building code amendments, shortcuts in building practices and violations that seriously undermined the integrity of the code and the quality of the building stock."

"A critical factor," the industry's report said, "is that there has not been progress in implementing hurricane-resistant construction, craftsmanship and inspection procedures along the coastline comparable to what has occurred in warning and evacuation procedures. In fact, progress toward such a goal often seems like two steps forward and two steps backward. Codes are improved and strengthened and then they are watered down under pressure to save building costs."[112]

Arguing against tighter codes

The leading sources of the pressure to save building costs by resisting stronger standards have been the National Association of HomeBuilders and local builders' associations. Along with insurers and academics, builders also have picked their way through hurricane debris, looking for answers to the questions on the minds of the victims whose houses failed. About Andrew's devastation, the NAHB said, "The winds exceeded the design specifications of all U.S. building codes and standards." The damage was so great, they seemed to say, because the hurricanes *were more powerful than anybody could have anticipated they would be.*

Saying—incorrectly—the wind in these costly hurricanes was unusually strong, the NAHB has continued to resist changes in building codes or building practices that they say would run up the cost of construction.

Raise the cost of a house by $1,000 and you'll find that 40,000 families fall out of the pool of potential new home customers, the NAHB warns. Rhetorically, the builders ask: Which is worse, for a family not to be able to afford a house at all, or for that family to worry that its insured house might be hurricane-damaged?

Reviewing the destruction caused by Andrew and Hugo led the homebuilders to say: "Simple but effectively applied plywood covering would have provided the needed protection" for many windows. They determined that in most homes surveyed, "Little regard was given to that kind of individual hurricane preparation."[113]

In other words, according to the NAHB, the storms were stronger than the design specifications for which the code was written, *and* homeowners' own negligence further contributed to the damage. The builders association was careful in its reports not to blame those who have written the codes nor the inspectors nor the developers and builders themselves.

To its credit, nevertheless, the National Association of Homebuilders did not try to excuse careless construction. The organization itself recommended new kinds of training and accountability "for all participants in the building process," presumably including carpenters, who tend to learn their trade from one another. It also recommended "improvements in building code requirements related to hurricane resistance."[114]

On the other hand, the builders have pushed hard for many years against wind-resistance standards that raise the price of their products.

Their message has the consistency of a drumbeat: "Cost must be a major consideration if building codes are to be changed."[115]

Even after the devastation of Andrew, the homebuilders strongly objected to code strengthening in south Florida, according to the former director of the National Hurricane Center, who served on a committee with them. "We got a few improvements in," Robert C. Sheets said, "but the builders quashed some things that certainly would have been beneficial to the public."

As for the homebuilders' influence pre-Andrew, *The Miami Herald* reviewed 35 years of minutes of Dade County's Board of Rules and Appeals. At least six times during that period, the board significantly lessened the code requirements in response to requests from builders for leniency or cost cutting, according to the newspaper.

Affordable hurricane resistance

Andrew's legacy since 1992 includes substantial changes in the Southern Florida's Building Code that the NAHB still argues "may be excessively costly relative to the benefit." Although the affordability argument has strength as a sound bite, two specific examples weaken it.

The first is a community known as Munne Estates in Dade County. The 71 houses there stood up well under Andrew's attack, in stark contrast to what happened to houses all around them. The strong houses were of concrete block and 5/8-inch plywood. The nails had been driven by hand. Continuous strips of mortar held on the roof tiles. The construction had been carefully supervised. The prices of the modest homes, with low-pitched but not flat roofs, ranged between $80,000 and $95,000.[116]

The second example is the fact that all 27 Habitat for Humanity homes in Dade County, including six in Homestead in the vicinity of Hurricane Andrew's highest winds were intact and habitable after the storm. One of them, built out of "in-steel," which is wire mesh with polystyrene and concrete blown over the top, came through unscathed in the midst of a neighborhood that was blown apart.[117]

Damage avoidable

Dr. Peter Sparks, Clemson University civil engineering professor, a man who has studied damage from every major hurricane for a decade, helped the insurance companies figure out what had changed their joyous period of policy-selling into the nightmarish period of claims-paying. Baloney to the notion that Hugo, Andrew and Iniki were so big and so bad that the destruction they delivered was inevitable, he said. Actually, all three hurricanes were typical in strength when compared to others that have hit the regions they hit.

"A lot of the hurricane damage could have been avoided," Sparks added. "Knowledge has not been translated into practice. These (hurricane) events are events we ought to have been able to resist."[118]

As a practical matter, even if a critical mass sprang up today and pushed successfully for new codes and improved enforcement of them, the nation's coastal communities still would have vast amounts of housing in place already that will fall apart in hurricanes. In 1990, in a random sampling of homes in Galveston, Texas, for example, Texas Tech University's Institute for Disaster Research found less than half in compliance with the local code. Except for National Flood Insurance Program's requirements, it also found *no* insurance companies giving incentives or making demands for the construction of wind and flood resistant homes.

Although the code is better today than ever before, according to the Texas Tech engineers, they nevertheless found "a great deal of residential property at risk in Galveston." They recommended training of designers, homebuilders and building inspectors to make the code effective. In addition, they recommended a series of retrofitting actions to reinforce the majority of the houses built before the new rules took effect.[119]

Engineers take up the problem

Insurance companies want to pay fewer and smaller claims for hurricane damage. The Federal Emergency Management Agency is trying to become proactive, hoping to prevent damage instead of merely responding to it. Looking for hard data to tell them how to reach their objectives, FEMA and the insurance industry funded research by Dr. Sparks, his colleagues and civil engineering graduate students at Clemson. They have studied designs and materials for the kinds of *houses individual families live in*, developing information that could benefit everyone who lives in the danger zone.

The engineers put various small models of houses, subdivisions and whole cities into a boundary layer wind tunnel, blow turbulent air over them and with sophisticated computer software study the pressure distribution. How great is the lift on the northeast corner of the roof? When will the windows blow out? Suppose this model of a house is set in an open field? Suppose it is set in a high-density downtown with a lot of trees around it?

How does a house "behave" in different wind conditions?

In addition to the precision-sensitive work in the wind tunnel, the academic engineers fire two-by-four pieces of lumber at targets—trying to duplicate what a hurricane does when it shakes the parts of a house loose and slams the parts into other houses. They test wood and roofing materials and nails and fasteners and other components.

Reviews of thousands of hurricane-damage claims convinced Sparks that " a small amount of wind damage creates an enormous amount of water damage." In short, if a storm shatters part of any building's "skin," the exterior walls and roofs that keep the wind and water outside, the structure will suffer. Maintain complete coverage and prevent big, inconvenient, perhaps even disastrous losses.

Commercial, industrial and multi-family buildings get structurally engineered frames—that usually stand up in all but the worst hurricane winds. However, many larger buildings in recent hurricanes have lost their non-engineered parts—the windows, doors, siding and roof coverings— and thus had their "skin" ripped. When that happens, the building's contents are wrecked, and the building is rendered unusable, despite its *structural* integrity.

The Clemson team describes a facetiously simple test to determine whether an oceanfront condominium complex will "make it" in a bad storm: If you can stick a pocket knife into the exterior wall, the answer is "no."

Measuring wind velocities

During most hurricanes, nobody actually records the wind velocities while the destruction is under way. Some anemometers are not accurate in high winds. The wind engineers who study the debris after storms complain that most of the time when hurricanes hit land, the most accurate, most expensive anemometers are hidden safely away to keep them from being damaged. Hurricane Andrew destroyed several anemometers in its path.

For lay people, including builders and building-supply manufacturers, compounding the problem of discussing wind velocities and protection against winds are variations in standards. To the National Hurricane Center, the phrase "sustained winds" means wind velocities consistent for one minute. To the World Meteorological Society, the phrase "sustained winds" means winds consistent for ten minutes. "Gusts" are something else altogether.

Inside offshore hurricanes, National Hurricane Center meteorologists now take wind measurements in a range between 90 feet and 45,000 feet high and use those velocities to *estimate* surface-level wind velocities. On the ground anemometers typically are set 33 feet high. The wind engineers insist that designing for true wind-resistance requires having more precise wind-speed data than they are getting now.

Once hurricane debris is lying all over the ground, everybody who looks at it begins to guess. How much power did it take to shake that oak tree's roots out of the ground? How fast was the wind that pushed that roof rafter through the palmetto tree? What wind velocity tore off the roof shingles? What velocity knocked the chimney down? Engineers apply their formula to the questions, but they can only estimate. Most of the time nobody knows the answers for sure.

A compilation of tips

Even without knowing all they would like to know about velocities, the Clemson engineers and others who have studied the problem nevertheless know a lot about holding small buildings together in high winds. From a half dozen sources comes the following composite of recommendations for house building in the hurricane zones:

- Instead of gable roofs, use hip roofs. Gable roofs invite the wind to suck off shingles and plywood. Hip roofs frustrate wind and thus assist in protecting the house's contents.

- Instead of low-pitched roofs, which generate uplift in strong winds, use moderate-pitched roofs.

- Instead of nailing down the roof sheathing, screw it to the rafters. Sheathing held by properly placed screws withstand five times more uplift pressure than sheathing by properly placed nails. To retrofit an existing house, spray a special foam that becomes glue along the seams between the sheathing and the rafters.

- Instead of mere nails for shingles and shingle adhesive, use nails and glue, or roof cement. Instead of lightweight shingles, use wind-resistant, heavy ones. Those testing to "60 miles per hour" are not strong enough. Whatever can be done to hold the roof covering on helps the cause.

- Instead of big, wide windows and glass patio doors, design smaller glass openings—no wider than three feet.

- Instead of tentatively perching uncovered vents on the roof, fasten them down tightly and cover them to keep the hurricane rainwater out of the attic. When a storm is headed toward them, many folk board up their windows but forget to batten down the vents on the roof.

- Instead of installing or leaving (in the case of existing structures) mechanical equipment such as heating and air condition systems and water heaters at ground level, elevate them above the anticipated flood levels.

- Instead of enjoying a yard filled with pine trees, likely to snap in high winds, plan a landscape that puts the big trees at least one tree length away from the house.

- Instead of allowing rotten or brittle trees to stand or lean near the house, prune regularly.

- Instead of windows without coverings, make sure windows have coverings, snugly fitted. Storm protection can be designed as ornamental shutters to match the house. They might be electrically driven systems. They might be five-eighths-inch plywood sheets, corrugated aluminum panels or plastic panels that can be bolted on while the hurricane chugs landward.

- Instead of double doors that can be easily blown open, install single doors. Make sure they are firm in their frames and that the frames that are firmly attached to the walls. Cover them when a hurricane is headed your way. In Hugo, the wind entered a discount store through the front door after the door had been rattled out of its flimsy frames—and thus was able to blow off the store's roof and thus destroy not only the building but hundreds of thousands of dollars worth of merchandise too.

- Instead of gently connecting the foundation to the ground, make sure every part of the house clings to every other part. The frame and walls should be bolted to the foundation, and the roof system should be strapped to the frame and walls.

All of the tips apply to manufactured housing (mobile homes) as well as "stick-built" houses, although many experts believe no manufactured housing is designed and constructed to withstand hurricanes.

When a house is under construction, incorporating tips for wind resistance costs no more than 5 to 10 percent and in some cases, less. A few inexpensive but strongly built houses held together against the 155-mph winds of Hurricane Andrew. Retrofitting existing buildings also usually can be done for 5 to 10 percent, depending. Is it worth the trouble? Only if the hurricane comes while you live in the house.

After it comes, those unprepared for it find themselves in a big mess. Sometimes the mess is so big and so severe they cannot crawl out from under it.

Those who have prepared their homes as thoroughly as possible find themselves in less of a mess. Everything around may be in shambles, but with a house to sleep in, anybody and any family is better off than the neighbors whose homes are smashed. For individuals, knowing how to keep a house intact during most hurricane conditions— and being willing to do it ahead of time—can make a world of difference.

Why don't more folk take steps to improve their odds for living happily ever after in the danger zone? In part, because they are attached to a long list of hurricane myths.

Chapter 8

Myths and Realities about Hurricanes

What you think you know can hurt you

"The trouble with people is not
That they don't know but that they know
So much that ain't so."

Josh Billings

Like hurricanes, dangerous myths about hurricanes can be organized into five categories. One category of myths offers unfounded opinions about hurricanes' behavior. A second category of myths suggests reliance on uninformed sources for critical information. A third category leads danger-zone residents astray on issues of property protection. A fourth category is based on the false assumption that hurricane victims have easy access to the resources needed to restore their homes and their communities. A fifth category of myths implies that the prophets of doom are just trying to make trouble.

Like hurricanes of all categories, hurricane myths of all categories are dangerous. For coastal residents, investors and hurricane-season vacationers, they lead to errors in judgment. Such errors making sitting targets of homes and businesses and the institutions that make communities hum. They have led to exorbitant damage costs in recent years. Such errors also can be matters of life and death. Although hurricane fatalities declined in the 20th century, experts expect them to rise in the 21st. The growing populations that continually increase evacuation times carry a myriad of illusions in their heads.

Hurricanes' behavior

Despite all that is known about hurricanes, they continue to harbor mysteries, which create ideal conditions for speculating and theorizing about what they will do. Much hurricane lore that people chat about is half true and therefore misleading. Dangerous pronouncements based on junk science include the following:

What you have to be afraid of is the high winds around the eye, so if the eye is not headed to you, you will be fine.

Indeed, it is true that the highest winds circulate around the lowest air pressure in the calm eye, but is also true that a hurricane's path of devastation can be surprisingly wide. Remember the Great Sea Island Storm of 1893 that laid waste a region in South Carolina almost as big as the state of Connecticut. Even more important is this truth: Far more people drown in hurricanes, either from the storm surge or rain flooding, than are killed by wind. This eye-wall myth is a recipe that could cost lives.

Hurricanes may be bigger than tornadoes, but at least they are predictable, so you can get out of their way.

That hurricanes will form in the Atlantic Ocean and the Gulf of Mexico when sea temperatures reach 80 degrees and will travel in a general westward and northward direction is predictable. However, as the National Hurricane Center's documented margins of error demonstrate, when and where and with what force they will strike land cannot be accurately forecast far enough in advance for every resident community to be guaranteed ample evacuation time.

Hurricanes mainly threaten the people on the beachfront and the barrier islands of the Atlantic and Gulf coastlines.

Typically, because of the storm surge, the closer to the ocean, the more damage. However, swaths of hurricane destruction often penetrate well beyond the coastal zones. As three examples of many, the Great Sea Island Storm of 1893 blasted farms and cities with wind and rain in a half dozen states, Hurricane Hugo in 1989 carried hurricane-force winds 150 miles inland and Hurricane Floyd flooded one-fourth of North Carolina in 1999.

We don't have to be as concerned about a little hurricane as a big one.

Other things being equal, a big hurricane will wreck a bigger region than a little one. However, size has nothing to do with intensity. Category 4 Hurricane Andrew in 1992, a beast in terms of total damage, was small in comparison to most hurricanes.

The same hurricane never strikes both Florida and Texas.

Don't count on it. The deadliest hurricane ever, the Galveston Hurricane of 1900, went through the Florida Keys on its way across the Gulf.

The shape of the shorelines of Georgia and Virginia somehow keeps hurricanes from coming into those states very much.

It is true that south Florida, coastal North Carolina and the Florida Panhandle have been hurricane struck more often than Georgia and Virginia in the last century, but hurricanes have devastated every section of the coastline from Maine to Texas. Go back to the risk probabilities listed near the end of Chapter 2, "959 Such Storms in 111 Years."

The cooler water east of the Gulf Stream causes the Atlantic hurricanes to turn north and pick up speed most of the time.

Sometimes, maybe, but not necessarily. Scientists believe the most useful data to help them forecast tracks are available in the top of the hurricane, about 50,000 feet above the surface, rather than in the ocean.

The friction caused by trees and buildings on the land cuts the wind.

That is not true. Hurricanes lose power over land because ocean waters under them no longer provide the heat that fires them and the moisture to sustain them. On the other hand, immediately after landfall, the wind-gust velocities increase around obstacles on the surface.

Hurricane-information sources

In the Information Age in an open society, Americans have the opportunity to be better informed about hurricanes than ever before. Accuracy, however, is not assured in everything that is published and broadcast. Almost anybody, regardless of training or lack of it, can look at a weather map and can go on the air or into Internet chat rooms with inaccurate information and foolish recommendations. Anyone who wishes also can dish out friendly, worthless advice in the checkout line at the supermarket. Misleading information such as the following can bombard you from dozens of sources:

The Internet is a great place to go for information because it gives you satellite and radar data. Ordinary people can look at the same data weather officials see and make their own judgments.

Yes, some Internet web sites provide frequent updates during a hurricane emergency, and some provide much raw data as well. Do not underestimate the complexity of hurricane forecasting, however. Lacking the skill, the tools and the professional experience to analyze the data, non-scientists can endanger lives with falsehoods and bad judgment.

TV meteorologists can be trusted because they understand better than the forecasters in their ivory towers what real people need to know during a hurricane threat.

TV meteorologists can and do provide critical information to the public during a crisis, but when they broadcast predictions and advice contrary to those of the National Hurricane Center, they confuse and mislead the public. Although the official forecasters do not get it right every time, their average is better than that of other forecasters. Only the National Hurricane Center has the advantage of multiple levels of review of the science and the technology, plus many years of experience, in preparation for issuing hurricane advisories. Besides that, understand that not all weather-news broadcasters are educated as meteorologists.

Locally elected officials know their communities best. The public should listen to them instead of listening to the state and federal bureaucrats.

Local officials certainly play a key role during a hurricane threat. Nevertheless, hurricanes approaching the continent usually threaten several hundred miles of coast at a time, so response to hurricanes usually must be coordinated among many jurisdictions. For the most effective results, local officials must cooperate with state officials, and officials of several states must coordinate with one another. On the matter of what to expect from the hurricane itself—its path, intensity, rain, flooding and timing—everyone should rely on information from the National Hurricane Center's forecasters only.

People who have lived on the coast for a long time know from experience how hurricanes behave; it is only smart to pay attention to what they have to say.

Listening to long-time coastal dwellers talk about storms is almost always entertaining and in many cases educational. Some can tell amazing, true stories about destruction and fatalities from certain hurricanes that have blasted their communities. And yet, some of them have worked out faulty rationale in their own minds for irrational decisions when a hurricane is spinning offshore. Although some coastal old-timers share their common sense based on experience with newcomers, others share their nonsense.

Hurricane forecasters and TV meteorologists make the hurricanes seem worse than they really are. They have to build up their ratings, and they don't want to take the heat if the worst happens.

Please understand that hurricanes sometimes intensify and change direction or forward speed unexpectedly and that the forecasters often do not know exactly what is going to happen next. The genuine hurricane experts know, though, that it is not in anybody's interest for them to cry, "Wolf!" They understand well that unnecessary fright and overreaction during one hurricane threat will backfire next time. And there is always a next time.

My house must be safe. Otherwise, they wouldn't have let the developer build it where it is. Also, the Realtor is a friend of mine, and I know she would have not have encouraged me to buy in a hazardous location.

Make no assumptions. The basis of coastal land-use planning is more often politics than science or sound investment advice. The basis of land development is economics—what will make money. Real estate licenses certify professionals to sell property but do not certify them as experts in hurricane safety. Contact your local emergency preparedness officer for information, preferably before buying. Go back to Chapter 7, "Did It Have to Be This Bad?" "No," to refresh yourself on these issues.

Notions about protecting property

The property-protection myths that have driven coastal development in the last half of the 20th century have set up many communities for catastrophe. Clear evidence of the problem stands in the exorbitant and rising costs of hurricane damage. That certain structures will one day become hurricane debris is guaranteed. The myths that in some cases leave residents and investors subject to wrenching losses include the following:

I have homeowners' insurance; I don't have to worry.

Homeowners' insurance does not pay for flood damage and may not cover the total cost of other damage. Unless your policy covers replacement of possessions important to you, you may be disappointed. If you live or own structures in a flood zone, you need flood insurance. If you live in a flood zone and your community is not one of the many communities in the National Flood Insurance Program, you should convince local officials to wise up. If you can't do that, move or run for public office yourself.

Building codes require hurricane clips and all kinds of reinforcements now. My house is new. It ought to be sturdy enough.

First, both complacency and resistance from builders' organizations have kept building codes from being as strong as they need to be in some regions. Second, builders do not always build according to codes in every detail, and building inspectors typically do not inspect every fastener and brace in a house. Third, many designs popular near the coast, and especially on waterfront properties, are inappropriate in high winds, but the building codes typically do not address these design features.

Well, my house has been here—thirty years, fifty years, one hundred years, whatever—and it's lasted through many a hurricane. I shouldn't have to worry about it.

Perhaps you are fortunate enough to own a home designed and built to withstand hurricanes. However, research by builders, insurance companies and wind engineers shows that most housing standing in the coastal zones today is likely to be significantly damaged when a hurricane comes.

Maybe you could build or retrofit houses to withstand hurricanes, but most people couldn't afford all that.

Actually, designing and building houses to hold together during the most common hurricane conditions is not cost prohibitive. While Hurricane Andrew demolished whole neighborhoods in Homestead, Florida, a few carefully constructed, modest priced homes held up well.

If a Category 4 or Category 5 hurricane barrels into your house, it's going to fly apart, and there's nothing you can do about it.

It may not be cost effective to build a house that will withstand a direct hit by the most powerful hurricane; however, coastal residents are more likely to experience Category 1, 2, and 3 hurricane conditions, and there is plenty that can be done to resist them.

Window shutters are a waste of time and money.

Actually, on a house or business with a strong roof, the difference between having a building left when the storm is over and not having one might well be determined by the soundness of the window and door coverings. Their value shows up when they keep the wind from prying its way inside. If you can keep the wind from coming in through the windows and doors, you may protect the interior from rain and you may protect the roof from the uplift that would blow it away.

I've always heard, "Open the windows on the downwind side of the house so the low air pressure of the hurricane doesn't cause the house to explode."

Yikes! Do everything you can to keep all windows and doors not only closed but also covered. Houses are not airtight balloons that will blow to bits because of the low air pressure in the hurricane.

Someone told me to put masking tape on the windows to keep them from shattering.

Masking tape will do nothing of value for windows subjected to hurricane-force winds and wind-driven debris. Great frustration may result in the effort to scrape the tape off the windows after the hurricane emergency is over.

Relief and recovery myths

Myths in the fourth category loom on the ocean's horizon as midsummer nights' dreams about what happens after a hurricane lays waste an entire coastal region. There is the perception that once a hurricane has wrecked a region, insurance payments, federal disaster declarations and massive volunteer assistance come in and put Humpty Dumpty back together again. Anyone who has experienced the aftermath knows better. Once the scenes of devastation, despair and anxiety disappear from the nation's TV screens, the survivors' traumatic predicaments disappear from the minds of the nation's audience. By a long shot, that doesn't mean the survivors' predicaments are over. Beware of the following widespread myths:

People respond with unbelievable kindness and generosity after a hurricane. Disaster brings out the best in human nature.

Many storm stories have in them heart-warming elements of heroism and generosity, and many levels of government, organizations and volunteers spring into action to help the victims of hurricanes. Unfortunately, such help is temporary, and storm survivors must fend for themselves in the long run. In addition to bringing out the best in human nature, disasters also bring out some of the worst. The unscrupulous know how to defraud folk under the stress of hurricane destruction, and some of them do. Recall that Charleston, South Carolina policemen sent to help Floridians after Hurricane Andrew arrested 95 looters within 12 hours after arrival in Dade County.

People who can't afford motels when they evacuate can always go to a shelter. The Red Cross takes care of them.

Take a reality check. First, there are never enough shelters to take care of all evacuees, and in some regions, the shortages are acute. Second, operated on donated dollars, shelters are selected on the basis of safety but are not equipped to be comfortable, convenient, pleasant or attractive. They do not take pets. Not only do shelter populations not have beds, sofas, TVs or VCRs; they often have long lines to the toilets and dining rooms. In addition, people under stress living in close quarters together, especially if they were strangers before, tend to annoy one another. Better to make other plans if at all possible.

Millions of dollars pour into areas damaged by hurricanes. People are better off after it's all over than they were before.

Insurance companies pay insurance claims, but a lot of property is either uninsured or underinsured. Repairing damage and rebuilding are fraught with difficulty and usually with higher than expected expenses. Where zoning laws have changed and where skilled workers are in great demand, people who live through a hurricane sometimes wonder whether they will make it through the stress of the aftermath. If a house or a business needs a new roof anyway, and the insurance company will pay for it after the hurricane winds ripped it off, OK. For a family even temporarily homeless because of a hurricane, though, the message that they will one day be "better off" offers thin comfort. For a business owner even temporarily unable to conduct business immediately after a hurricane, the downtime can be impossible to overcome.

We all pay taxes. The government gives grants and makes low-interest loans to help people. It's just a matter of standing in line to fill out the forms.

For most of the government's financial assistance programs, only those with no resources or meager resources of their own are eligible. Those who do qualify for loans have to pay them back, usually under difficult circumstances.

What will be, will be. If the storm destroys my business, I'll just build it back.

Reality is not on your side. Most businesses need a healthy, intact community in which to thrive and most hurricane-ravaged communities do not thrive again normally for at least several years after the damage. Surveys show that about 50 percent of the businesses hit by a hurricane either do not rebuild or do not succeed if and when they do rebuild.

The 'prophets of doom'

Myths in the fifth category, arising out of human nature's desire to be happy and not worry, take the prize for being the most insidious of all. They encourage danger-zone residents to ridicule those who warn them. They discourage precautionary action of all kinds. The fantasies listed below are based on the old habit of "denial," which supposedly justifies neglect of hurricane preparations. The belief that everything will turn out all right in the end is expressed in at least the following ways:

I'm not going to spend my life planning for a hurricane that may never come. There are disasters: tornadoes, mud slides, blizzards, volcanoes, wildfires, crimes, something. Life is risky. You can go out on a highway and get killed, so what's the use of worrying about hurricanes?

Wherever savvy folk live, they make adjustments for the peculiar risks in their environments. Different risks require different precautions. Most drivers, for example, do what they can to minimize their risks on the highway. Coastal dwellers who bring judgment to their knowledge about hurricanes can take reasonable steps to reduce risk to their lives and property. They are not hard to do.

You've got to outwit those "masters of disaster" who call for an evacuation every time they get a chance. I might leave but I am going to wait till the traffic congestion is over, look at the hurricane on the weather map and then zip on out if it's coming and stay right where I am if it's not.

In a full-blown evacuation, by the time the traffic has cleared out, it may be too late for whoever is left on the coast to flee before the winds pick up and the roads flood. Also, evacuation of a crowded coastal region can fill up every public shelter space, every hotel and every motel for as much as 200 miles inland. Except for trying to ride out a hurricane in a beach cottage or a mobile home, trying to ride it out in a car is the worst of all choices.

I'm in the tourism business on the coast. I can't afford to shut down everything because of a "false evacuation" in the peak of the season. That takes money out of my pocket. When I have to pay the price for somebody else's mistake—calling for an evacuation when the hurricane is going somewhere else— somebody's head is going to roll.

Whether or not to call for an evacuation is tricky for elected officials, but smart business people treat evacuation costs as a part of doing business on the coast. As a result of population growth in the danger zone, evacuations will occur more frequently in the future. Rolling the heads of those who order evacuations will not remedy the problem—which is that hurricane forecasting accuracy is not keeping up with coastal populations.

I've wanted to live near the coast all my life. Now that I'm about to retire, you're telling me it's a "danger zone"! Forget it. I have a right to be near the beaches, and I'm willing to take my chances.

Taking chances is a choice anybody might want to make. Our concern is for those who do not realize the chances they are taking when they live in the most vulnerable coastal regions. As a retiree, you might want to consider that destruction of a home and family belongings is tough to overcome, and tougher for the elderly than the young. You also might want to think about the tiring, tedious nature of typical hurricane evacuations.

Good-sounding but deceptive

Many of the hurricane myths carry the ring of folk wisdom. It's tempting to listen to them, to repeat them, even to believe them. They satisfy a natural desire to latch on to easy responses to troublesome truths. The sad fact is that many of the myths are espoused by folk who do not know what they are talking about.

Sound-bite sayings, as appealing as they may be, obscure the complexity of the hurricane problem. By proffering misleading answers to difficult problems, they cheat those who believe them. They stymie thought and stymie action. In order to move toward solutions for themselves and their communities, danger zone residents have to discard the myths. Danger zone residents who work at looking beyond the myths to discover the truths of their predicament will find bad news at first: The whole problem is more complicated than it appears at first glance. Fear and dread are legitimate reactions to the hurricane vulnerability of millions of Americans.

Danger zone residents who continue to explore the situation will come upon good news eventually: As big and complex as the problem is, it can be broken into parts, which then can be sorted and addressed. Optimism has a legitimate place, too, if attitude adjustments lead to sensible responses to the realities of living and investing where hurricanes ravage the landscape.

Chapter 9

What Can be Done about it?

Plenty

> *"The great end of life is not knowledge but action."*
>
> **Thomas Henry Huxley**

If coastal dwellers could bury the myths, and keep them buried, they could reduce the dangers to themselves and their communities. Individuals, families and businesses can take certain actions in their own best interests. In addition, those willing to become activists have the power to initiate movements for massive changes that affect whole regions.

For everyone who lives within fifty miles of the coast from Maine to Texas, two simple site-specific investigations are well worth the time they will take. Doing them will be interesting and fun.

The first is this: Experience an adventure by finding out how hurricanes have treated your region in the past. Every Atlantic and Gulf Coast region has a hurricane history worth exploring.

Old-timers, old newspapers, the National Weather Service's records and local libraries are repositories of hard-to-believe facts and personal recollections about hurricanes. Along with the data on wind, water, damage and loss of lives, chase down the telling anecdotes. Given the unforgettable nature of hurricane experiences, drama is guaranteed. One at a time, human stories make their contribution to the complex narrative of each storm.

About recent hurricanes, expect to discover tidbits such as this: *The evacuation traffic was so ungodly it took 20 hours to drive 200 miles, there was nowhere to stop and spend the night and nowhere to go to the bathroom.*

About hurricanes anywhere at any time, you might find these gems of information: *When the water finally drained off, someone found a huge fish in the Catholic church. A man drowned trying to save his shrimp trawler. A yacht came to rest on its side in the middle of Lockwood Avenue. A boxcar landed in the river. The storm ripped the roofs from 54 stores in a single supermarket chain before it was all over. They found a golden retriever beside the crib in the roofless house down the street.*

The second investigation can be equally enlightening. Call on your local emergency management office and your nearest American Red Cross chapter to learn about their hurricane preparations. They will be glad somebody wants to know what they are doing. Ask for information from the most recent flooding-prediction SLOSH (Sea, Lake and Overland Surge from Hurricanes) study and for estimated evacuation times under various conditions. Find out if and when water will invade the properties important to you.

Ask to read or be given summaries of the agencies' pre-hurricane and post-hurricane planning documents. Inquire about the frequency of updates and mock-disaster rehearsals. Who is in charge of making sure the communication channels function in the crisis? Who decides whether to evacuate or not?

Once you become aware of your community's hurricane history and its current hurricane-response program, you will be ready to start on your mission. Combine your information and the ideas you have with what you learn from FEMA (Federal Emergency Management Agency) and the

Red Cross. They provide ample how-to information in booklets, brochures and web sites. For presentations to groups, they have slide shows and videos.

After studying the FEMA and Red Cross material on hurricanes, develop your own detailed, personalized "to do" lists. Brainstorm with and enlist the commitment of those close to you who will be most affected by the hurricane threat or the hurricane's arrival. You will need a "to do" list for your home and your family. You will need another for your business. You may need one for organizations for which you are responsible, including nonprofit institutions such as churches. Take the following recommendations into account as you prepare your own program for action:

Before hurricane season begins

- Arrange for a qualified professional to check the hurricane-readiness of structures important to you, including your own dwelling, your business, your mother's nursing home, your church, or whatever you value. Correct what needs correcting.

- Buy or build window coverings and devise a system for installing them, either as a permanent part of the structures you want to protect, or as temporary shutters to install when a hurricane threatens. Make realistic provisions for shuttering the windows above the first story.

- List and photograph your significant belongings. Make sure your property insurance covers flooding and wind damage to pay for whatever you are not willing to lose. Store your important papers and small, valuable, irreplaceable items in a waterproof vault or in a safety deposit box in a bank—a bank outside of the flood zone.

- Plan ahead for business disruptions, for pet protection, for ways to handle special obligations unique to your family or your neighborhood. Your disabled and elderly family members and friends are probably going to need help if they have to evacuate and even more help if a hurricane makes landfall near your community.

- As the season nears, settle evacuation plans; make arrangements for storage of vehicles, boats, horses and computers during a hurricane; prune trees; gather and store basic supplies, including a first-aid kit, nonperishable foods, batteries, matches and tools. Identify, list and plan how you will pack the small, transportable valuables for hurricane evacuations.

During a hurricane threat

- When hurricane forms offshore, tune in to media sources frequently, keep the vehicles fueled, get basic information from family members and key associates, including employees, about their plans. Take the hurricane into account in all major decision-making.

- Make up your mind to rely on forecasts originating in the National Hurricane Center and follow the recommendations of state and local emergency officials, including recommendations of early voluntary evacuations for those who can leave ahead of the masses.

- If the hurricane appears headed toward you, relocate boats, vehicles, horses and computers.

- Make sure your supply of medications, including prescriptions, is sufficient for a couple of weeks.

- Nail down evacuation plans with reservations at motels, state parks, families or friends living away from the danger zone or with directions to a specific shelter; collect last-minute foods and drinking water and medicines.

- Store a supply of potable water in a safe place in your home or business.

- When the hurricane is imminent, start installing window and door coverings far enough ahead of time to finish the installations before getting an anticipated order to evacuate. You may not be able to hire anyone to do such tasks for you in the critical hours. Clear the yard of anything that will become a projectile in storm winds. Take care of your commitments for protecting other people and facilities for which you are responsible.

- If forecasters become confident your community is not going to be struck, and if you do not have to evacuate, wait and watch for a few hours just to be sure. If the hurricane really misses you, put things back as they were and rejoice at your good luck. Think of the exercise you have just completed as a "dry run." Use what you learned this time for the next time.

- If you must evacuate, touch base at the last minute with family members and associates about their evacuation plans. Pack one vehicle only, cover the door through which you exit and then get onto the highway, traveling toward a safe haven on high ground. Take a map. Go deliberately into a relaxed mode. Assume that traffic will be heavy and painfully slow in places. Expect to take two to three to four times as long to travel as would be necessary under normal circumstances.

- Along with your insurance documents, your computer back-ups, your portable radio and your house plans (if available), take books on tape or music, take joke and riddle books, take snacks and drinks. For children, take games and lap toys. Ban whining from the vehicle.

- If you are going to a public shelter, take pillows, sleeping bags, medications, snacks and reading material. Do not take pets or alcoholic beverages.

- If you do have to evacuate, understand and do not fret about this fact: The hurricane is likely to go elsewhere.

During and after a hurricane

- If the hurricane slams your community, take a deep breath and face the truth that the situation may not be "normal" again for a long time. Understand and do not fret about the fact that you may not be able to return to your home immediately. Public authorities have responsibilities for public health and safety. Hurricane demolition leaves the place in a hazardous condition. You do not want to compound the hurricane disaster with avoidable accidents.

- Once you get back to your home, avoid dangers such as contaminated water, downed live wires, weakened structures and poisonous snakes that may have taken refuge in your house. Check on family and friends; determine how best to provide food and shelter for at least a few days; identify the closest source of information and help; begin simple repairs and clean-up. Call your insurance agent.

- If you cannot avoid staying in the danger zone during a hurricane, stay inside, preferably in a room without windows and outside doors, and do not be fooled by the calm eye. Identify your next of kin in case officials must notify someone of your demise.

Something else to consider

If you are unable or unwilling to cope with hurricane preparations, you should live outside of the danger zone. Otherwise, you may face dangers you will not like, and you may also become a problem for others.

If you are a retiree thinking of moving to the coast, recognize that the older you get, the harder it will be to handle the physically taxing tasks required to protect yourself and your property. Imagine yourself in a few years trying to install window shutters and door coverings and trying to clear the yard of everything the hurricane may pick up and use for battering. Imagine yourself driving long hours to safety or trying to sleep on the floor of a schoolhouse corridor. Imagine yourself "starting over" after a hurricane has ripped through your home.

What others could do

Along with doing what must be done as individuals, coastal dwellers can rally their neighbors. Improving the degree of safety in the danger zone will require seismic changes in institutions as well as in the habits of those who live there. In some enlightened communities, institutional changes are under way. In most, only the power of the people, the taxpayers and the consumers can exert enough influence to make real the shift in culture necessary to effect the following recommendations:

- Insurance companies could require stronger building rules as a condition of issuing insurance policies in the danger zone.

- Architects could design houses and other buildings to resist wind damage. The designs could include provisions for hurricane shutters.

- Community developers could limit the creation of new homes, businesses, streets and sewerage systems on flood-prone land. On land that is safe most of the time but expected to be under water as a result of major hurricanes, they could publicize the risk factors along with the lots' selling prices.

- Engineers and builders could utilize wind-resistant designs and construction practices in coastal regions—and train construction crews to carry them out. Maybe one day they will promote their services on the basis of such capabilities.

- Local governments could adopt land-use laws that limit development in the low-lying, waterfront lands, especially on the islands. They could restrict the pace of development permitting to the pace at which they can raise highway capacity to handle the traffic new development generates, including hurricane-evacuation traffic.

- Governments also could publicize risk assessments and locations of the flood-prone areas—and adopt sufficiently strict building codes. They could make sure the building-inspection teams are appropriately trained, staffed and supervised in order to enforce the codes.

- Government officials, law officers, National Guardsmen and others involved in handling a hurricane emergency could be equipped to communicate with one another.

- Every county and municipality in the danger zone could develop a post-hurricane *recovery* plan, complete with budgets, contracts and permits. They also could develop a post-hurricane *redevelopment* plan, complete with ordinances and funding sources ready to be put to use.

- Every coastal state, county and municipality could make training in hurricane-disaster management a priority for its employees. To learn what goes on elsewhere as a result of hurricanes, emergency personnel should participate in the hurricane-planning and hurricane-response experiences of communities other than their own.

- The states could prohibit construction of new nursing homes and major hospitals in the areas that will be flooded by hurricanes. (This prohibition need not apply to emergency medical facilities.)

- State and local officials could update sequences, routes and maps for evacuation annually and could make sure the public knows about them before hurricane season, as well as during a hurricane threat.

- The state highway departments could shoulder proudly their responsibility for public safety in hurricane evacuations and cooperate at every level to facilitate traffic flow at critical times. In some regions, they should be ready on short notice to implement a plan of lane reversal, so that evacuees can use three lanes of a four-lane highway for traffic to get away from the coast.

- Congressional representatives from the coastal states could become champions of the National Hurricane Center and the Federal Emergency Management Agency, making sure they get the resources needed to deal with these life-and-death issues.

- Congress could revamp the National Flood Insurance Program so that it will more effectively discourage growth in the flood-prone zones of highest risk.

- Candidates for public office in the danger zones could be familiar with their communities' evacuation plans, zoning and building codes and the steps to be taken in the aftermath of a hurricane. Voters should reject the uninformed.

- The American Red Cross could spotlight the problems of the shortages of hurricane shelters and work with coastal communities' private and public sectors to address those shortages. A variety of solutions may be necessary.

Since those who live outside of the danger zone are not going to take up the banner of hurricane protection, it is up to the coastal dwellers to learn what is needed, to muster the courage to fight for policy changes and to persevere. Even a single helpful change on the part of a key player might trigger changes on the part of others. Toward the goal of cultural shift as well as the objective of improvements in public policies, the coastal dwellers of the 18 vulnerable states could petition their legislators and governors. The language in the draft resolution in the appendix could serve as a model for petition documents. A single motivated organization could organize a coalition of coastal organizations to take this approach.

Tougher issues to tackle also

It might take a series of hurricane catastrophes to motivate danger-zone residents to begin demanding dramatic shifts in the way things are done where they live. And yet, it is hard to believe that coastal dwellers will sit silently forever. When they begin to speak out on the importance of planning for hurricanes, however, they will meet opposition they probably never expected.

Powerful forces hold the thorny problems firmly in place.

Start with the culture of land development. The tidal wave of people and money moving to the coast shoves aside everything in its way. Unmanaged growth is treated like the source of all prosperity and all that is good about this country. In many states, the every-man-for-himself version of property rights is a mantra. In the coastal regions dominated by growth-and-development at all costs, hurricane-related facts are treated like family scandals, sometimes acknowledged privately but unmentioned in public.

A second force sure to resist meaningful changes is the normal reality of the way coastal community leaders set priorities. Taxpayers and voters make multiple demands on mayors, town councils and county governing bodies. They expect schools, police, ambulances, ball fields and beach accesses. They require courts and hospitals. Hurricane preparedness rarely rises to the top of most coastal politicians' priority lists. Requiring study, thought, cooperation, change, money and commitment, typically hurricane preparedness is a politically unpopular cause when the sun is shining.

Then there is the gambler's instinct. Three to four coastal areas are struck every year by hurricanes or tropical storms, but nobody can say further than several hours ahead of time which areas they will be. For even the most hurricane-prone sections of the danger zone, in any given year, the likelihood of suffering from a major hurricane is low. It is not natural for most of us to think of drownings, of mud and water on the kitchen floor, of long lines to get soup or ice and of dislocation for months or even years as tragedies that will happen to ourselves. So, blow on the dice and roll them one more time.

Difficult quagmires

Taken together, the forces that encourage everybody to stick to the present course are almost overwhelming. Combine those forces with the unwieldy nature of some of the problems. As time goes on, everything grows more complex and more intractable.

For example, if you are familiar with a rapidly growing or densely developed coastal community, think about and talk to your friends about what is needed to relieve the traffic congestion. The drawbacks to significant improvements will be obvious. Most states' highway systems need serious upgrading on a statewide basis, and there is no political consensus that coastal areas should be at the front of the line to receive the necessary tax dollars. Except in communities that have evacuated recently, the coastal public does not connect the dots between the forecast uncertainties, the evacuation timing dilemmas and highway-capacity issues. And officials who are not from the coast spend precious little time pondering how the next hurricane evacuation will occur.

In many coastal communities, building new and more effective evacuation routes would require condemnation and demolition of expensive property. Some would require construction of long, exorbitantly priced bridges—if environmental permits could be obtained. Some would require the erection of overpasses and other upheavals in the landscape that would be costly, disruptive and for other reasons objectionable.

Clearly, despite the importance of rapid evacuations, most of the coastal public cannot expect miraculous new highways to get them out of the jam they are in.

Assessing the difficulties of emptying the danger zone quickly will force coastal folk to face the lack of safe sanctuaries near the coast. Although most hurricane evacuees prefer not going to public shelters, because of such long evacuation clearance times, some may have no choice. When all the hotels and motels within 150 miles are filled and the hurricane is on its way, people might be forced to do what they do not want to do. However, there are far too few safe shelter spaces to accommodate everybody who might need one.

More public shelters nearer where the endangered populations live would reduce the evacuation times for some and could save lives. The facilities of colleges, schools and courthouses, maybe even the facilities of military installations, in nearby inland counties, for example, could be used in emergencies if they were designed and built—or retrofitted—to resist most hurricanes. Major questions and significant objections to those ideas arise immediately. Who would negotiate such uses? Who would pay the cost? How would that old bugaboo of legal liabilities be addressed?

Considering the need for near-coast shelters raises the need for emergency hurricane shelters in the danger zone itself. It is a topic nobody likes. In some regions, because of lengthy evacuation clearance times, where not everybody can evacuate, there are only two possible safety valves to relieve the situation, both hazardous and undesirable.

Danger-zone residents and visitors might go upstairs in a multi-story building and be better off during a hurricane than they would be at ground level. If such escapes are planned for ahead of time, they are called "vertical evacuations." If the evacuees are fortunate, they are supplied with camp toilets, military type rations, bottled water and flashlights. Although the drawbacks to vertical evacuation are substantial, it is an answer more attractive than unplanned-for "last-resort sheltering."

Last-resort sheltering comes into play at the last minute, as winds shriek and waters rise, and people scamper into buildings that they hope will protect them. There are no plans ahead of time to accommodate those who scamper under such conditions. If people are lucky in these last-resort places, they will not be hit by flying glass or falling roofs; however, they will have no access to supplies or services. Understandably, emergency officials hate the concept of last-resort sheltering because they do not want to tell the public to run into a "shelter" that may be unsafe.

Among the hazards of such attempts to escape danger is the simple truth that so many buildings in the coastal regions have been neither designed nor constructed to withstand the worst forces hurricanes can deliver. Some that have been appropriately engineered and are structurally sound in many cases have exterior walls, doors and windows that will fail.

As scary and difficult as the vertical-evacuation possibilities and last-resort options are, political leaders in certain coastal regions are obligated to work on them. Along with the American Red Cross and other shelter providers, they should call private enterprise and many jurisdictions together to brainstorm for solutions, develop plans and reach innovative agreements in the interest of saving the lives of their constituents.

During some of this nation's hurricanes of the past, danger-zone residents scrambled onto whatever was higher to keep from drowning. Rooftops and tree limbs saved lives in the Great Sea Island Storm of 1893 and in the Galveston Hurricane in 1900. In some of the danger zone's most crowded communities now, unless extreme measures are taken, in extreme hurricanes, thousands of people could find themselves unable to flee from a charging storm surge or rising flood waters. That high numbers of Americans might drown in a hurricane in this century seems unthinkable, but the unthinkable appears to be poised for reality.

In most jurisdictions, chaos when a hurricane comes is a certainty. Except in the rare places in which intense hurricane preparations have been made, neither the governments nor coastal infrastructure nor the economy nor the people are ready for the prime time event of a hurricane landfall. And yet, hurricanes come, ready or not.

Research, enlightenment and progress

On what, then, can coastal dwellers hang their hope?

Stunned by the costs of recent disasters of many kinds, the Federal Emergency Management Agency in the 1990s started promoting "mitigation"—a comprehensive program of action taken in advance to minimize losses, costs and miseries. It has enlisted 200 communities and more than 1,000 businesses in its Project Impact.

For the hurricane-prone communities, FEMA knows what needs to be done—by individuals, businesses, nonprofit organizations and governments. The program addresses building construction, emergency communication and debris management, among many other issues. Comprehensive and time-consuming to execute because many, many groups and jurisdictions have to be involved, Project Impact nevertheless provides guidance, as well as funds, to communities wanting to take a giant step in the right direction. For more information on Project Impact, including a presentation to assist groups willing to brainstorm about the possibilities in their hometown, visit the FEMA web site.

Another good sign is that the state of Florida—the state with the most exposure and with a fresh memory of the catastrophe a small, intense hurricane (Andrew 1992) can create—has upgraded its preparedness efforts substantially since the early 1990s. Surcharges on property insurance premiums are earmarked for emergency planning. Evacuation routes are posted on everything from community bulletin boards to grocery bags. Land in Florida's most flood-prone zones now has restrictions on intensity of use. Florida is growing at the rate of 1,000 persons per day, and Florida already has in place millions of people and billions of dollars worth of buildings that cannot withstand hurricanes. So the risks remain. Still, other states would do well to copy some of what Florida has done as a result of Hurricane Andrew.

In addition, weather scientists and technology experts are spending time and energy on the problems of forecasting and evacuating. As one of many examples, Charles Watson's storm-prediction model TAOS (The Arbiter of Storms) and other refinements for analysis of data are coming along to supplement the flooding-prediction model SLOSH (Sea, Lake and Overland Surge from Hurricanes), the official tool of the National Hurricane Center. The International Hurricane Center is pushing for aerial surveying with laser beams and for more precise warnings about rainfall risks. The nation's official forecasters are working with the International Hurricane Center on a new hurricane-impact model to complement the familiar Saffir-Simpson Hurricane Scale. Such advances offer promise in the quest to become more selective in deciding who must evacuate in various scenarios. The important objective is to clear out everybody subject to the storm surge. If others in the danger zone stayed in appropriately built and prepared houses or other buildings, they would not enjoy the hurricane experience, but at least they would not drown in ocean water, and they would have a chance of making it through OK. Communications before and during the crisis would be critical, but the selective evacuations in some regions might be effective in saving lives.

"We cannot continue to evacuate as many people as we do. We have to have a tool to help fine-tune these evacuations," said Stephen Leatherman, director of the International Hurricane Center in Miami.[120]

Max Mayfield, who became director of the National Hurricane Center shortly before the 2000 hurricane season, recommends storm shutters, "safe rooms" and other steps to prepare homes against hurricanes. "Safe rooms" would not work for buildings standing in the flood zones but could be effective where wind is the most likely culprit. "You flee from storm surge, you hide from wind," Mayfield said.[121]

The professionals in weather and in emergency management understand best the high-stakes gamble in the hurricane-vulnerable regions.

They know they have a responsibility for public safety, and many of them do all they can to improve the odds in favor of protecting lives and property. They all hope that a hurricane disaster that takes a lot of lives will not occur on their watches.

And yet, the professionals have a limited ability to save populations that have settled where hurricanes have come and gone for centuries. The people who can really make a difference are the people with the most to lose. Real solutions will begin to move into place only when those who are living in the danger zone get fired up and take action on their own behalf.

About the authors

Both William F. "Bill" Marscher II and Frances "Fran" Heyward Marscher, husband and wife, grew up in Beaufort County, South Carolina, a community clearly in the danger zone—and they live there now.

Bill was educated as an engineer at Clemson College and earned masters' degree from the Massachusetts Institute of Technology. He worked for MIT on the Apollo moon program before returning to Beaufort County in 1969 to become a vice president of the Sea Pines Company on Hilton Head Island. In that capacity, he participated in regional environmental initiatives. Since then, he has owned and operated a building supply business, sold commercial real estate and retired to spend time on state and local public service and writing projects.

Fran was educated as a journalist at the University of South Carolina. From 1987 until January 1997, when she retired, Fran was editor of The Island Packet, Hilton Head Island's daily newspaper, a property of McClatchy Newspapers Inc. As a reporter earlier, she won numerous writing awards. As editor, she and the news staff won national honors.

Bill and Fran have spent three years doing research for and writing *Living in the Danger Zone* and *The Great Sea Island Storm of 1893*. Both are active in local and state public service organizations.

The Marschers do, indeed, install window shutters when a hurricane threatens their home. They also grow vegetables and herbs in raised beds year-round.

Appendix

Saffir-Simpson Hurricane Intensity Scale				
(Based on National Hurricane Center information)				
Saffir-Simpson Category	Maximum Sustained Wind Speed (miles per hour)	Barometric Pressure (inches of mercury)	Storm Surge Height (feet)	Potential Damage
1	74-96	greater than 28.93 (typical sea level is 29.93)	3-5	**Minimal**. No real damage to building structures; most damage to unanchored mobile homes, trees and signs; coastal road flooding and minor flood damage; some small craft torn from moorings.
2	97-111	28.93-28.50	5-8	*Moderate. The potential wind damage of a 111- mph category 2 is 1.3 times greater than that of a 96-mph category 1.* Some damage to roofing, doors and windows; considerable damage to mobile homes, trees and signs; considerable damage to piers; marinas flooded; some small craft torn from moorings; coast roads and low-lying evacuation routes cut by rising water; evacuation of some shoreline residences required.

Saffir-Simpson Category	Maximum Sustained Wind Speed (miles per hour)	Barometric Pressure (inches of mercury)	Storm Surge Height (feet)	Potential Damage
3	112-131	28.50-27.90	8-12	*Extensive. The potential wind damage of a 131-mph category 3 is 1.9 times greater than that of a 96-mph category 1* Structural damage to some buildings; mobile homes destroyed; coastal structures damaged by flood debris; substantial regional flooding extends along rivers and sounds; more evacuation routes cut by rising water; evacuation of more shoreline residences required.
4	132-155	27.90-27.17	12-18 (Hurricane Hugo:20.8)	*Extreme. The potential wind damage of a 155-mph category 4 is 2.6 times greater than that of a 96-mph category 1.* Extensive structural damage; some complete roof failures; major damage to lower floors of structures near shore; all terrain lower than ten feet above sea level may be flooded; massive evacuation of residential areas within six miles of shoreline required.
5	greater than 156 (Hurricane Camille estimated at 195)	less than 27.17 (Hurricane Gilbert 26.3)	Greater than 18 (highest recorded: 34 in Australia)	*Catastrophic. The potential wind damage of a 195-mph category 5 is 4.1 times greater than that of a 96-mph category 1.*Complete roof failures on many residences and commercial buildings; some complete building failures; major damage to all structures located less than 15 feet above sea level; massive evacuation of all residents within 10 miles of shoreline required.

Hurricanes' Names

Hurricanes' names provide handy reference points for discussion purposes. For example, Mississippians refer to periods in the state's history as "before Camille" and "after Camille."

Federal meteorologists gave hurricanes women's names from 1953 to 1979, when they started using men's names as well. The list below is recycled every six years, except that names of hurricanes that go down in infamy because of the destruction they cause are "retired" and replaced.

2000	2001	2002	2003	2004	2005
Alberto	Allison	Arthur	Ana	Alex	Arlene
Beryl	Barry	Bertha	Bill	Bonnie	Bret
Chris	Chantal	Cristobal	Claudette	Charley	Cindy
Debby	Dean	Dolly	Danny	Danielle	Dennis
Ernesto	Erin	Edouard	Erika	Earl	Emily
Florence	Felix	Fay	Fabian	Frances	Franklin
Gordon	Gabrielle	Gustav	Grace	Gaston	Gert
Helene	Humberto	Hanna	Henri	Hermine	Harvey
Isaac	Iris	Isidore	Isabel	Ivan	Irene
Joyce	Jerry	Josephine	Juan	Jeanne	Jose
Keith	Karen	Kyle	Kate	Karl	Latrina

2000	2001	2002	2003	2004	2005
Leslie	Lorenzo	Lili	Larry	Lisa	Lee
Michael	Michelle	Marco	Mindy	Matthew	Maria
Nadine	Noel	Nana	Nicholas	Nicole	Nate
Oscar	Olga	Omar	Odette	Otto	Ophelia
Patty	Pablo	Paloma	Peter	Paula	Philippe
Rafael	Rebekah	Rene	Rose	Richard	Rita
Sandy	Sebastien	Sally	Sam	Shary	Stan
Tony	Tanya	Teddy	Teresa	Tomas	Tammy
Valerie	Van	Vicky	Victor	Virginie	Vince
William	Wendy	Wilfred	Wanda	Walter	Wilma

Sample resolution

For maximum preparedness against the disasters hurricanes can deliver, coastal residents should take a series of essential steps personally, and then stir up others to act also. To make significant progress toward the safety of their communities, they must create hurricane awareness among their neighbors and with their neighbors push businesses and governments toward hurricane-smart policy changes.

Although the complexity of the problem makes it difficult to figure out where to start, hurricane-relevant information from local emergency management staffs and regional American Red Cross headquarters provides ammunition for an attack on the status quo. One early tactic in the battle for comprehensive improvements could be circulation and presentation of a petition to elected officials, modeled on the following draft:

Whereas, hurricanes are the most destructive forces in the planet's weather system, with winds up to more than 155 mph and storm surges ranging to 20 feet deep;

Whereas, coastal *(your state)* is likely to be hit by a Category 1 or 2 hurricane every *(number based on National Hurricane Center's risk assessment)* years and by a Category 3, 4 or 5 every *(number based on the Hurricane Center's risk assessment)* years;

Whereas, the *(certain hurricane or hurricanes of note in your state's history)* demonstrated the power of hurricanes to cause severe, long-term suffering among the stricken victims;

Whereas, as many as *(estimated permanent population plus vacationers)* might be living and vacationing in this state's coastal zone on any given day for the hurricane season, June through November;

Whereas, typically, most of the dwellings in the coastal regions are not built to hurricane-resistant standards;

Whereas, the National Hurricane Center's documented margin of error in forecasting means that 200 miles of coastline face severe risks and many uncertainties 24 hours in advance of a hurricane landfall;

Whereas, the number of people who will have to flee periodically from the coast to prevent possible death from hurricane drowning is steadily growing;

Whereas, the state's transportation network lacks the capacity to handle present hurricane evacuation traffic in a timely manner;

Whereas, evacuation times in coastal *(your state)* range from *(number of hours in best-case scenario to number of hours in worst case scenario)*, and evacuation should be done in daylight if at all possible;

Whereas, the more hours it takes for a coastal region to evacuate, the earlier evacuation decisions must be made during a hurricane threat;

Whereas, earlier decision-making means that evacuations will occur frequently when the hurricane goes elsewhere as well as when the hurricane hits;

Whereas, evacuation in more than one state at a time can create traffic gridlock and chaos within a matter of hours;

Whereas, inside a vehicles is one of the most dangerous places to be during a hurricane;

Whereas, the misery of a single arduous evacuation might dissuade populations from trying to evacuate next time a hurricane threatens.

Therefore, the State of *(your state)*, in conjunction with local governing bodies, is obligated to evaluate this very serious risk to its population and take actions in the interest of human safety and the common welfare.

Therefore, a *(your state)* State Blue Ribbon Commission on Hurricanes shall be appointed immediately. The appointees should include specialists in building construction, land-use planning, traffic management, political science, hurricane meteorology, disaster recovery, law enforcement, property insurance and public communications as well as emergency management experts. The commission should be chaired by an individual familiar with the issues, committed to information gathering and capable of building consensus. Appropriate state resources, including staff, should be assigned to work with the commission.

The commission should evaluate situations and trends that affect the state's ability to withstand hurricanes and hurricane threats, including but not limited to:

- The carrying capacity of coastal highways.
- The location of housing and commerce in the storm-surge zones.
- The pace of coastal population increases.
- The design and the quality of construction in the areas subject to storm surge and in the areas subject to hurricane-force winds.
- The potential for sufficient evacuation shelters within 75 miles of the coast.
- The capability of the numerous hospitals, nursing homes and assisted-living facilities in the state's hurricane-evacuation regions to execute evacuations.
- The level of coastal residents' knowledge about hurricanes and hurricane preparedness.
- The quality of every coastal community's post-hurricane plan.

The commission then should make two "to do" lists: one that can be addressed quickly and a second that will require long-term commitments. The recommendations should be directed to those in authority at all levels to change whatever is required in the interest of public safety and the common welfare.

Recommended reading

Academic papers

Ayscue, Jon K., "Hurricane Damage to Residential Structures: Risk and Mitigation," November 1996, The Johns Hopkins University.

Platt, Rutherford H., and Dawson, Alexandra D., "The Takings Issue and the Regulation of Hazardous Areas," June 1995, University of Colorado Natural Hazards Research and Applications Information Center.

Agency publications

American Red Cross and U.S. Department of Commerce, National Oceanic and Atmospheric Administration, "Hurricanes, The Greatest Storms on Earth," March 1994.

Federal Emergency Management Agency, "Coastline at Risk: The Hurricane Threat to the Gulf and Atlantic States," 1992.

Institute for Business and Home Safety, "Is Your Home Protected from Hurricane Disaster?" 1997.

Institute for Business and Home Safety (formerly Insurance Institute for Property Loss Reduction), Coastal Exposure and Community Protection Hurricane Andrew's Legacy."

U.S. Environmental Protection Agency, Office of Solid Waste, "Planning for Disaster Debris," December 1995.

Books

Barton, Clara, *The Red Cross in Peace and War*, American Historical Press, 1903.

Barnes, Jay, *North Carolina's Hurricane History*, University of North Carolina Press 1997.

Barnes, Jay and Frank, Neil, *Florida's Hurricane History*, UNC Press 1998.

Bush, David M., Pilkey, Orrin H. Jr., and Neal, William J., *Living by the Rules of the Sea*, Duke University Press 1996.

Dunn, Gordon E. and Miller, Banner I., *Atlantic Hurricanes*, Louisiana State University Press 1964.

Hanson, Gunnar, *Islands at the Edge of Time*, Island Press/Shearwater Press 1994.

Larson, Erik, *Isaac's Storm: A Man, A time and the Deadliest Hurricane in History*, Crown Publishers 1999.

Pielke, Roger A., *The Hurricane*, Routledge, London and New York, 1990.

Pielke, Roger A. Jr., and Pielke, Roger A. Sr., *Hurricanes: Their Nature and Impacts on Society*, John Wiley & Sons Inc., 1998.

Savadors, Larry and Buchholz, Margaret Thomas, *Great Storms of the Jersey Shore*, Down the Shore Publishing 1997.

Sill, Benjamin L. and Sparks, Peter R., editors, *Hurricane Hugo One Year Later*, American Society of Civil Engineers, 1990.

Tannehill, Ivan Ray, *Hurricanes*, Princeton University Press 1952.

Booklets

Hurricane Survival Guide, Staff of the Sun-Sentinel, Fort Lauderdale, Florida 1993.

Internet

National Hurricane Center, www.nhc.noaa.gov

Federal Emergency Management Agency, www.fema.gov

American Red Cross, www.redcross.org

The Weather Channel, www.weather.com

International Hurricane Center, www.ihc.fiu.edu

Periodicals

Fortune, "A New Way to Bet on Disasters," September 1997.

Kiplinger's Personal Finance Magazine, "Yikes!" November 1996.

Disasters: The Journal of Disaster Studies and Management, "Protecting Tourists from Death and Injury in Coastal Storms," 1996.

Weatherwise, "The Intensity Problem," September/October 1998.

Glossary

"Alphabet soup": slang for the multitude of agencies with acronyms common in emergency preparedness and disaster relief.

Clearance time: the estimated number of hours between the moment the public is notified of a decision to evacuate and the moment the last family is out of the evacuation zone.

Cyclone: low atmospheric pressure characterized by circulating winds.

Declaration of disaster: Presidential announcement that severe life and safety conditions in a certain region warrant special assistance from the federal government. The help comes in the forms of services, grants and loans.

DAC: *Disaster Application Center*, official headquarters for FEMA's processing of requests for assistance. DACs are established in federally declared disaster areas.

EOC: *emergency operations center*, headquarters (in some jurisdictions) of persons handling official functions during a community crisis.

EPD: *emergency preparedness division*, headquarters (in some jurisdictions) for agency responsible for handling situations hazardous to the public.

EMA: *emergency management agency*, headquarters in some jurisdictions for agency responsible for handling situations hazardous to the public.

EVACUATION: procedure for removing people from danger zones. Under hurricane threats, evacuations typically affect waterfront properties, properties subject to flooding from extreme rains and mobile homes.

FEMA: *Federal Emergency Management Agency*, founded in 1979. The name says it. The head of FEMA answers to the president. FEMA coordinates preparation for disasters, takes the lead in getting many federal agencies involved in disaster relief.

FEMA posts 800 numbers for victims to call and register their needs.

Hurricane: Tropical storm in the North Atlantic, in the Northeast Pacific Ocean east of the dateline and in the South Pacific Ocean east of 160 degrees east, with sustained winds in rotary circulation of 74 mph (called a typhoon in the Northwest Pacific Ocean west of the dateline).

Hurricane watch: threat of hurricane conditions within 24-36 hours somewhere along the coast.

Hurricane warning: threat of hurricane conditions within 24 hours or less in a specified coastal area. Actions for protection of life and property should begin immediately.

Intensity: description of severity of a hurricane, based on winds and air pressure and summarized in the Saffir-Simpson Scale.

International Hurricane Center: non-profit hurricane research organization created in 1997 and situated on the campus of Florida International University in Miami.

Maximum sustained surface winds: Average wind speeds. The National Hurricane Center (NHC) and the Joint Typhoon Warning Center (JTWC) of the USA average wind speeds over a one-minute period. The World Meteorology Organization averages wind speeds over a 10-minute period.

Mitigation: process of taking steps ahead of time to minimize adverse impacts.

National Hurricane Center: Headquarters of the U.S. Weather Service's hurricane-related operations, under the National Oceanic and Atmospheric Administration, located on the campus of Florida International University campus in Miami.

SBA: Federal Small Business Administration, which makes low-interest loans to businesses and individuals under certain conditions during a disaster recovery period.

SLOSH: Sea, Lake and Overland Surges from Hurricanes, a numerical computer model used for predicting storm surge. SLOSH studies form the basis for most hurricane planning in this country.

Storm surge: large dome of water 50 to 100 miles wide that sweeps across the coastline near where a hurricane eye makes landfall. The stronger the hurricane and the shallower the offshore water, the higher the surge will be. The area of greatest storm surge impact is in the right forward quadrant of the hurricane.

TAOS: The Arbiter of Storms, a relatively new method of modeling the effects of large scale storm systems, using numerous data sources including satellite imagery, public domain information and meteorological parameters.

Tracking models: mathematical computer models used to track intensity and track of tropical cyclones.

Statistical models forecast the future by using current information about the hurricane and comparing it to historical knowledge about the behavior or similar tropical cyclones.

Dynamical models use global atmospheric data on wind, temperature, pressure and humidity to make forecasts based on current conditions.

Combination models combine historical data with current data to try to capitalize on the strengths of the other two systems.

Tropical cyclone: low-pressure system over tropical or subtropical waters with thunderstorm activity and definite cyclonic wind circulation.

Tropical depression: tropical cyclone with sustained surface winds in rotary circulation of 38 mph.

Tropical storm: tropical cyclone with winds in rotary circulation of 39-73 mph sustained winds, typically named.

Typhoon: Tropical storm in the Northwest Pacific Ocean west of the dateline, with sustained winds in rotary circulation of 74 mph (called "hurricane" in the North Atlantic Ocean, in the Northeast Pacific Ocean east of the dateline and in the South Pacific Ocean east of 160 east).

Wave run-up: Phenomenon of waves running ahead of incoming surf. Wave run-up may occur inland of the storm surge.

Wave setup: Process that causes still-water surfaces to incline upward locally to elevations above the prevailing sea level as deep-water waves move into rapidly shoaling area. Wave setups occur on top of storm surges, and waves occur on top of wave setups.

Notes

preface

[1]www.redcross.org

[2]American Meteorological Society and The Weather Channel Inc., September 2000, "Policy Issues in Hurricane Preparedness and Response."

Chapter 1

[3]U.S. Department of Commerce, National Oceanic and Atmospheric Administration, "Hurricane," 1971.

[4]www.aoml.noaa.gov/hrd/tcfaq

[5]Owen Ullmann, Paul Overberg and Rick Hampson, Growth reshapes coasts; a wave of development overwhelms the shore," *USA Today*, July 21, 2000.

[6]David M. Ludlum, *Early American Hurricanes* (American Meteorological Society, 1963).

[7]Ibid.

[8]Ibid.

[9]U.S. Census.

Chapter 2

[10]Herbert Molloy Mason, *Death from the Sea* (The Dial Press, 1972)

[11]Ibid.

[12]Ibid.

[13]Ibid.

[14]"Andrew! Savagery from the Sea," *Fort Lauderdale* (Florida) *Sun-Sentinel* (Tribune publishing, 1992).

[15]Ibid.

[16]Ibid.

[17]Ibid.

[18]Jay Barnes, *North Carolina's Hurricane History* (University of North Carolina Press, 1995).

[19]Leslie H. Tyler, *The New England Hurricane, an album of pictures* (City Printing Company, New Haven, Connecticut, 1938).

[20]Ibid.

[21]Ibid.

[22]*The Devastation and Restoration of New England's Life Line*, a publication of The New Haven Railroad, Boston.

[23]www.nhc.noaa.gov, "tropical cyclone records."

[24]Insurance Research Council and Institute for Property Loss Reduction, *Coastal Exposure and Community Protection, Hurricane Andrew's Legacy*, 1995.

[25]Wallace Kaufmann and Orrin Pilkey, Jr., *The Beaches Are Moving* (Duke University Press, 1979).

[26]Claudia Smith Brinson, *Hugo!* (State Printing Company, 1989).

[27]Insurance Research Council Inc. and Institute for Property Loss Reduction.

[28]www.redcross.org

[29] Robert Sheets, former director of the National Hurricane Center, "United States Hurricane Problem: An Assessment for the 1990s, "14th American Hurricane Conference, 1992.

Chapter 3

[30] Insurance Research Council Inc. and Institute for Property Loss Reduction.

[31] *USA Today*, July 21, 2000.

[32] Ibid.

[33][34]Insurance Research Council Inc. and Institute for Property Loss Reduction.

[35]Robert C. Sheets, "The United States' Hurricane Problem."

[36]Chris Coudriet, hurricane program planner, state of North Carolina, personal correspondence with authors September 1997.

[37]Insurance Research Council Inc. and Institute for Property Loss Reduction.

[38]Watson Technical Consulting, interviews of Charles Watson by the authors.

Chapter 4

[39]www.fl.state.ys/fdhc/pio/press/nhupdt4.html

[40]Robert C. Sheets, "America's Hurricane Problem: An Assessment for the 1990s," citing Ocean City, Maryland., Chamber of Commerce, 1992, p. 2

[41]Coudriet.

[42]Anonymous source, interview June 1997.

[43]Jerry Jarrell, former director of National Hurricane Center, interview with authors January 1998.

[44]Joe Modicut, Environmental Services Coordinator, Louisiana Department of Transportation, interview with authors August 1999.

[45]Donald Lewis, senior associate, Post, Buckley, Schuh and Jernigan Inc., "Evacuation Clearance Times for a Much Larger Population at Risk," National Hurricane Conference 1992.

[46]Greater Atlantic City Chamber of Commerce, correspondence with authors August 1997.

[47]Lewis, 1992.

[48]Ibid.

[49]Ibid.

[50] "So many people—and nowhere left to run," *USA Today*, July 25, 2000.

[51] Ibid.

[52]Amy Hughes, Savannah Area Manufacturers' Council director, interview with authors March 1997.

[53]Ibid.

[54]Ibid.

[55]Jay Baker, associate professor, Department of Geography, Florida State University, Georgia Hurricane Conference speaker, June 1997.

[56]Max Mayfield, National Hurricane Center director, presentation for South Carolina Hurricane Conference 1998.

[57]Ibid.

[58]Ibid.

[59]Joe Modicut.

[60]*Miami Herald* Sept. 19, 1999.

[61]Robert R. Collins, Hurricane Preparedness, Department of Community Affairs, state of Florida, correspondence with authors, September 1997.

[62]Ibid.

[63]Jimmy Chandler, attorney, interview with authors, spring 1998.

[64]Federal Emergency Management Agency, Florida Department of Community Affairs, and Lee County (Florida) Division of Public Safety, *Stormwatch! Hurricane Preparedness for Hospitals!* video, 1995.

[65] Ibid.

[66]Ibid.

[67]Pat Goodale, American Red Cross disaster specialist, interview with authors, fall 1997.

[68]Michael Logan, national American Red Cross Hurricane Coordinator, interview with authors, fall 1997.

[69]Robert C. Sheets, interview with authors, 1998.

[70]Charles Chesnutt, U.S. Army Corps of Engineers, interview with authors, 1998.

Chapter 5

[71]Jerry Jarrell, interview with authors ,1998.

[72]Robert C. sheets, *National Hurricane Center —Past, Present and Future*, 1990.

[73]Ibid.

[74]Ibid.

[75]Ibid.

[76]Ibid.

[77]Ibid.

[78]Ibid.

[79]Ibid.

[80]Jarrell, interview with authors, January 1998; Lawrence, Miles, National Hurricane Center forecaster, presentation to National Hurricane Conference, April 1997.

[81]Jarrell.

[82]Jarvinen, Brian, storm-surge specialist, National Hurricane Center, interview with authors, June 1997.

[83]Donald C. Lewis, interview 1998; Charles Chesnutt, interview 1998.

Chapter 6

[84]Pat Goodale, American Red Cross disaster specialist, interview with authors 1997.

[85]Southern Baptist Association officials, interviews with authors 1998.

[86]Claire B. Rubin. and Roy Popkin, "Disaster Recovery after Hurricane in South Carolina," Center for International Science, Technology and Public Policy, George Washington University, 1990.

[87]Rubin and Popkin.

[88]Ibid.

[89]Ibid.

[90]Ibid.

[91]Ibid.

[92]Robert Collins, correspondence with authors 1997.

[93]U.S. Army Corps of Engineers and Federal Emergency Management Agency, 1993, *Hurricane Andrew Assessment—Florida,*

[94]*Path of Destruction—Hurricane Andrew*, Historic Publications, Charleston, SC 29413 1992.

[95]Bobby Parker, editor, *The Savage Season*, Wilmington Star-News Inc., Wilmington, NC 1996.

[96]Ibid.

[97]Coudriet.

[98]The Associated Press September 23, 1999.

Chapter 7

[99] *USA Today*, July 21, 2000.

[100] American Meteorological Society and The Weather Channel Inc., September 2000, "Policy Issues in Hurricane Preparedness and Response."

[101]U.S. Environmental Protection Agency, Office of Solid Waste,

1995, *Planning for Disaster Debris.*

[102]Insurance Research Council and Institute for Property Loss Reduction.

[103]J. Smith Harrison, S.C. Wind and Hail Underwriting Association, interview 1997.

[104]*Andrew! Savagery from the Sea.*

[105]Ibid.

[106]Insurance Research Council and Institute for Property Loss Reduction.

[107]Ibid.

[108]Ibid.

[109]Institute for Home and Business Safety, successor to Institute for Property Loss Reduction, interview 1999.

[110]Insurance Research Council. Inc. and Institute for Property Loss Reduction.

[111] Ibid.

[112] American Meteorological Society and The Weather Channel Inc.

[113] Insurance Research Council and Institute for Property Loss Reduction.

[114]U.S. Department of Housing and Urban Development, NAHB Research Center, 1993, *Assessment of Damage to Single-Family Homes Caused by Hurricanes Andrew and Iniki.*

[115]Ibid.

[116]Ibid.

[117]Insurance Research Council and Institute for Property Loss

Reduction

[118]Ibid.

[119]Peter Sparks, Civil Engineering Department, Clemson University, interviews with authors 1997 and 1998.

[120]James R. McDonald and Billy Manning, *Effectiveness of Building Codes and Construction Practice in Reducing Hurricane Damage to Nonengineered Construction*, Institute for Disaster Research, Texas Tech University 1990.

Chapter 9

[121]Stephen Leatherman, Executive Director, International Hurricane Center, interview with authors January 1998.

[122] Associated Press, May 11, 2000.